Pop Music Production

Pop Music Production delves into academic depths around the culture, the business, the songwriting, and most importantly, the pop music production process. Phil Harding balances autobiographical discussion of events and relationships with academic analysis to offer poignant points on the value of pure popular music, particularly in relation to BoyBands and how creative pop production and songwriting teams function.

Included here are practical resources, such as recording studio equipment lists, producer business deal examples and a 12-step mixing technique, where Harding expands upon previously released material to explain how 'Stay Another Day' by East 17 changed his approach to mixing forever. However, it is important to note that Harding almost downplays his involvement in his career. At no point is he center stage; he humbly discusses his position within the greater scheme of events. *Pop Music Production* offers cutting-edge analysis of a genre rarely afforded academic attention.

This book is aimed at lecturers and students in the subject fields of Music Production, Audio Engineering, Music Technology, Popular Songwriting Studies and Popular Music Culture. It is suitable for all levels of study from FE students through to PhD researchers. *Pop Music Production* is also designed as a follow-up to Harding's first book *PWL from the Factory Floor* (2010, Cherry Red Books), a memoir of his time working with 1980s pop production and songwriting powerhouse, Stock Aitken Waterman, at PWL Studios.

Dr Phil Harding started at the Marquee Studios, London, engineering for acts such as The Clash, Killing Joke, Toyah and Matt Bianco. During the 1980s, Phil was chief mix engineer for Stock Aitken Waterman (PWL Studios), working with acts such as Dead or Alive, Mel & Kim, Bananarama, Rick Astley, Pet Shop Boys and Kylie Minogue. In the 1990s, Phil and Ian Curnow ran P&E Music Studios from The Strongroom Studio complex, producing hits for East 17 (including 'Stay Another Day'), Deuce, Boyzone and 911. Phil's first book, *PWL from the Factory Floor* (2010, Cherry Red Books), was followed by productions for Holly Johnson, Samantha Fox, Belinda Carlisle and Curiosity. He completed his doctorate in Music Production at Leeds Beckett University in April 2017.

Mike Collins began his music recording career in 1981 as a songwriter-producer at Chappell Music Publishing, where he achieved chart success with leading Brit-Funk band Light of the World. As part of a vibrant songwriter's workshop in the mid-1980s, Mike co-wrote with a range of established writers and artists such as Sly Fox, Cameo, Shannon, Jermaine Stewart and Joyce Sims. This led to top-flight session work from the late 1980s as a studio musician and MIDI programer working with Cubase and Pro Tools. Throughout the 1990s and 2000s Mike developed a successful technology-driven writing career with a highly respected series of books for Focal Press (Routledge) on Pro Tools. Mike leaves us with excellent music productions and his Pro Tools books, which are stocked in most university and college libraries.

Perspectives on Music Production

This series collects detailed and experientially informed considerations of record production from a multitude of perspectives, by authors working in a wide array of academic, creative and professional contexts. We solicit the perspectives of scholars of every disciplinary stripe, alongside recordists and recording musicians themselves, to provide a fully comprehensive analytic point-of-view on each component stage of music production. Each volume in the series thus focuses directly on a distinct stage of music production, from pre-production through recording (audio engineering), mixing, mastering, to marketing and promotions.

Series Editors

Russ Hepworth-Sawyer, York St John University, UK

Jay Hodgson, Western University, Ontario, Canada

Mark Marrington, York St John University, UK

Mixing Music
Edited by Russ Hepworth-Sawyer and Jay Hodgson

Audio Mastering: The Artists
Discussions from Pre-Production to Mastering
Edited by Russ Hepworth-Sawyer and Jay Hodgson

Producing Music
Edited by Russ Hepworth-Sawyer, Jay Hodgson and Mark Marrington

Innovation in Music
Performance, Production, Technology, and Business
Edited by Russ Hepworth-Sawyer, Jay Hodgson, Justin Paterson, and Rob Toulson

Pop Music Production
Manufactured Pop and BoyBands of the 1990s
Dr Phil Harding
Edited by Mike Collins

For more information about this series, please visit: www.routledge.com/ Perspectives-on-Music-Production/book-series/POMP

Pop Music Production
Manufactured Pop and BoyBands of the 1990s

Dr Phil Harding
Edited by Mike Collins

Routledge
Taylor & Francis Group

NEW YORK AND LONDON

First published 2020
by Routledge
52 Vanderbilt Avenue, New York, NY 10017

and by Routledge
2 Park Square, Milton Park, Abingdon, Oxon, OX14 4RN

Routledge is an imprint of the Taylor & Francis Group, an informa business

Library of Congress Cataloging-in-Publication Data
Names: Harding, Phil, 1957– author. | Collins, Mike, 1949– editor.
Title: Pop music production : manufactured pop and boybands of the 1990s /
Dr Phil Harding ; edited by Mike Collins.
Description: New York, NY : Routledge, 2019. |
Series: Perspectives on music production | Includes bibliographical references and index.
Identifiers: LCCN 2019010011 (print) | LCCN 2019012004 (ebook) |
ISBN 9781351189798 (Master e-book) | ISBN 9781351189781 (Pdf) |
ISBN 9781351189767 (Mobi) | ISBN 9781351189774 (ePub3) |
ISBN 9780815392804 (hbk : alk. paper) | ISBN 9780815392811 (pbk : alk. paper)
Subjects: LCSH: Popular music—1991–2000—History and criticism. |
Popular music—Production and direction—History. | Boy bands—History.
Classification: LCC ML3470 (ebook) | LCC ML3470 .H366 2019 (print) |
DDC 781.6409/049—dc23
LC record available at https://lccn.loc.gov/2019010011

ISBN: 978-0-8153-9280-4 (hbk)
ISBN: 978-0-8153-9281-1 (pbk)
ISBN: 978-1-351-18979-8 (ebk)

Typeset in Times New Roman
by codeMantra

Dedicated to my wonderful and supportive family
Frances, Jenny and James

And in Memory of
Mike Collins

Contents

Illustrations

Author and Editor Biographies

Dr Phil Harding

Phil Harding joined the music industry in 1973 at the Marquee Studios, London, engineering for the likes of The Clash, Killing Joke, Toyah and Matt Bianco by the late 1970s.

In the 1980s, Phil mixed for Stock Aitken Waterman tracks such as 'You Spin Me Round' by Dead or Alive followed by records for Mel & Kim, Bananarama, Rick Astley, Depeche Mode, Erasure, Pet Shop Boys and Kylie Minogue.

In the 1990s, Phil set up his own facility at The Strongroom with Ian Curnow. Further hits followed with productions for East 17 (including 'Stay Another Day'), Deuce, Boyzone, 911 and Let Loose.

Recent projects include the book *PWL from the Factory Floor* (2010, Cherry Red Books) and mixing Sir Cliff Richard's 2011 album Soulicious. Harding has recently worked for Holly Johnson (Frankie Goes to Hollywood), Tina Charles, Samantha Fox, Belinda Carlisle and Curiosity with his new production team PJS Music Productions. Phil is also Co-Chairman of JAMES (Joint Audio Media Education Services) and was the Chairman of the Music Producers Guild in 2003. Phil completed his doctorate in Music Production at Leeds Beckett University, April 2017.

Mike Collins, Editor

Mike Collins began his music-recording career in 1981 as a songwriter-producer at EMI Records and Chappell Music publishers, where he achieved early chart success with leading Brit-Funk band, Light of the World. 'Ride The Love Train' by Light of the World, co-produced and co-written by Mike Collins, reached number 40 in the UK charts in November 1981.

As part of a vibrant songwriter's workshop in the mid-1980s, led by legendary producer Gus Dudgeon at Dick James Music publishers, Mike co-wrote with a range of established writers and artists. Several years of TV work, producing recordings mostly for visiting US pop, soul and R&B artists in the 1980s, put Mike in the producer's chair with Sly Fox, Cameo, Shannon, Jermaine Stewart, Joyce Sims and similar acts. In 1987 Mike was in-house Senior Recording Engineer & Music Technology Specialist at the Yamaha R&D Studio in London W1. This brought opportunities to work with producers George Martin and Gus Dudgeon, rock musician

Keith Emerson, jazzman Courtney Pine and composers Hans Zimmer and George Benjamin.

This led to top-flight session work from the late 1980s onwards as a studio musician and MIDI programer working on singles for The Style Council, Zeke Manyika, Patsy Kensit, The Chimes, Eurythmics, Jeremy Healy, Feargal Sharkey and others. As soon as the new generation of digital audio recording and editing tools became available, Mike was an early adopter, using Cubase Audio and an early version of Pro Tools.

The 1990s saw Mike continuing his production, engineering and musician work for many artists in his own London-based recording studio. It also saw Mike's first foray into writing a series of in-depth books for Focal Press (Routledge) on Pro Tools users and thus began a successful technology-driven writing career. Since 2000, a move back to music production, Pro Tools engineering and co-songwriting led to work on several albums ranging from pop (Phil Harding's 'The Story of Beginners' album) and rock (the self-titled 'Catcha' album) to folk/pop (Caro's 'The 4th Way' album).

Between April 2013 and March 2014 Mike took a year out from music production to write two new books about Pro Tools 11 and all the latest audio plug-ins and virtual instruments; getting 'up-to-speed' with all the latest features along the way. Mike leaves us with excellent music productions and his highly respected series of Pro Tools books, which are stocked in most university and college libraries around the world.

Foreword

By Russ Hepworth-Sawyer

To be asked to write a foreword for Phil Harding's book is an honor I never expected. This is particularly important to me as much of my fascination with popular music, and what sparked my career, came from the records I listened to in the 1970s and also the transition to polished pop production in the 1980s. Phil, of course, made a significant contribution to both eras, and in particular the sound in the 1980s, through his recording and mix engineering as the chief engineer for the Stock Aitken Waterman production partnership. The SAW team had success with hits such as 'I Should Be So Lucky' by Kylie Minogue, 'Never Gonna To Give You Up' by Rick Astley and 'You Spin Me Round' by Dead or Alive, amongst many others. For someone truly interested in the 1980s and 1990s transformation of pop music, Phil's contribution to the engineered 'polished' sound, alongside Hugh Padgham (for example), should not be underestimated.

When Jay Hodgson and I set about to create the Perspectives on Music Production series for Routledge we were clear about its main objective: To publish relevant texts that had, where possible, both academic and industrial representation. This book is perhaps the epitome of these objectives as Phil Harding reflects on his career beyond SAW and his considerable success with the BoyBand culture of the 1990s.

Harding uses this forum to venture much further than to present a pure autobiographical reflection of his time and experiences with his writing partner, Ian Curnow, as P&E Productions. He delves into academic depths around the culture, the business, the songwriting and most importantly, the music production process. In doing so, Phil analyzes, retrospectively, the systems adopted, whether consciously or subconsciously. He explores the roles within a polished pop production team and continues this ethos with his current production team, PJS Music Productions.

Phil makes some poignant points, namely, that pure popular music is rarely cherished or valued enough to generate the level of academic attention that this book affords, in this instance towards BoyBands. This record therefore is more important to the literature on the subject and will become, I'm certain, a highly-cited document in the field.

Phil balances autobiographical discussion of events and relationships with academic analysis of how creative teams function, and the all-important *'flow'* that he has retrospectively observed in his and Ian Curnow's prolific work in the 1990s. Similarly, it is fascinating how Phil has listed the equipment used at the time and the technological barriers P&E were battling with, using beta-tested versions of Cubase, or pushing the limits of MIDI generated sounds to create lush productions for artists such as East 17. However, it is important to note that Phil almost downplays his involvement in his career. At no point is he center stage. Instead, Phil humbly discusses his position within the greater scheme of events. The closest we get to this book's story being about Phil is in his 12-step mixing technique, with which he expands on previously released material to explain how 'Stay Another Day' by East 17 changed his approach to mixing forever.

As any academic in the field of popular music or music production will tell you, the field is still relatively young. Therefore, to be in a position to analyze cultural shifts, production process techniques and popular music compositional methods in detail and all first hand adds weight and depth to the authenticity of the material available to you here.

Preface

For readers of my first book, *PWL From The Factory Floor* (Cherry Red Books, 2010), this book can be considered a follow-up, but unlike my first book, this one is not a memoir, it is predominantly an academic text. In 2014 I decided to reflect upon my career from 1992 onwards and was honored and humbled to be given permission by Leeds Beckett University to write about this period as the core of my thesis for a PhD by Portfolio in Music Production. The full title of the PhD was *'Stay Another Day': A Reflective and Oral History of The Technology and Culture of Manufactured Pop and BoyBands of the 1990s.* This book is derived from my PhD and includes up-to-date versions of some theories and pop music formulae that I discovered throughout my research and reflective practice. Much of the book details my working practices with my production and songwriting partner, Ian Curnow, at our P&E Music Studios, London, throughout the 1990s in a style that twentieth-century philosopher, Husserl, describes as the 'lived experience' (Husserl, 1927).

The musical genre that P&E focused on throughout the 1990s was the BoyBand and manufactured pop scene. Manufactured pop had dominated the UK charts since the mid-1980s, largely led by the Stock Aitkin Waterman (SAW) pop music songwriting and production team at PWL Studios, but the new pop music phenomenon in the 1990s was the rise of BoyBand popularity and culture. The first BoyBand to arrive on the scene in the early 1990s was New Kids On The Block (NKOTB) from the USA and hot on their heels were British BoyBands, Take That and East 17. Ian and I worked on early material by both of these acts but our path of destiny for the rest of the decade would be set by our newly-appointed manager, Tom Watkins, the music business entrepreneur who had guided the Pet Shop Boys and Bros to fame and fortune in the 1980s with his Massive Management team. Tom felt that we could not be seen to be working with both BoyBands, who quickly became rivals, his view was that we had to make a choice and that choice was easily made due to the fact that Tom was managing both East 17 and Ian Curnow and I (P&E). That decision led to a colorful and successful three-year working relationship between P&E/Tom Watkins and East 17, plus many other BoyBands and manufactured pop acts throughout the 1990s.

Acknowledgements

I would firstly like to thank my wife, Frances, for her constant support and encouragement throughout my doctoral thesis work and through to the completion of this book. I would also like to thank my wonderful interview respondents: Tom Watkins, Ian Curnow, Matthew Lindsay, Nicky Graham and Deni Lew, Trevor 'Tee' Green and John McLaughlin.

My heartfelt gratitude goes out to my friends and colleagues at Leeds Beckett University for their PhD guidance and encouragement from 2014 onwards: Dr Steve Parker, Dr Bob Davis, Professor Karl Spracklen, Dr Paul Thompson, et al.

Finally, a huge thanks to Russ Hepworth-Sawyer for making this publication available through the 'Perspectives On Music Production' Routledge series and everybody at Routledge for their guidance. Last, but not least, Mike Collins, my supportive and dedicated editor.

Illustrations

Thanks to Vince Canning at Coastline Graphics Ltd (www.coastline. com) for his excellent work on Figures 4.1, 6.5, 8.1, A.1 and A.2.

Photographs

Front cover photo courtesy of Mike Banks:
www.recordproduction.com
Back cover photo: Stan Shaffer, 2008.

Introduction

Methodology: Research and Interview Questions

The domain of pop music production and songwriting has been dynamically shifting since the 1950s and is often defined by the media (television, radio and music press) in terms of decades. In some instances in the UK, radio stations now even name themselves with reference to musical decades, for example, 'Absolute 80s'. There have been suggestions in popular music culture of how pop song and production formulas might be duplicated (Cauty & Drummond, 1988). There are also system model suggestions from studies on creativity (Csikszentmihalyi, 1992, 1997) and the musicology of music production (Zagorski-Thomas, 2014) describing the interaction within the domain, the field and the individuals involved in creativity, discovery and invention. However, human interaction within the group creativity involved in a recording studio and turning creativity into a repeatable commerce is relatively unexplored and leads to the question: Was there a formula or system framework behind the phenomenology of the 1990s BoyBands?

This book explores the shared culture of the 1990s manufactured pop and BoyBands, containing supporting evidence from my portfolio of work as a producer and composer in the 1990s with successful British artists such as East 17, Deuce and Boyzone. Interview participants, including managers, producers and songwriters, offer their reflective views and experiences of the culture and technology used to create this phenomenon in pop history. Interpretative phenomenological analysis was used to catalogue the interview data into the study areas. The research introduces an empirically and theoretically grounded formula framework proposal 'Service Model for (Pop Music) Creativity and Commerce' that started in the 1980s and can be contextualized today with ethnographic reflection. Observational case studies were used to highlight that the compositional and production elements of a successful pop and BoyBand system framework are intertwined. Evidence also suggests that for individuals to successfully participate in the field and domain of pop and BoyBand music they will need to display a genuine creative 'love' of the genre's culture and technology. Here, I bring all of these processes, systems and historical studies together within the book so that current and future practitioners in the pop music production field may have some useful start-point references.

The BoyBand phenomenon of the 1990s is set in a period that followed a successful decade of manufactured pop records in the 1980s, largely produced in the UK and Europe. This led to a narrow vision by record companies and artist managers as to what type of music young teenagers and pre-teens were likely to purchase. Data for 1980s manufactured pop is referenced from my published work, *PWL From The Factory Floor* (Harding, 2010). However, very little has been written critically about how manufactured pop records are created. Detailed accounts and perspectives from producers in various music genres of what happens in recording studio control rooms have been considered (Massey, 1994; Burgess, 1994, 2014). The music genre that I call 'manufactured pop and BoyBands' is more specialized than most other pop and rock music, which incorporates a variety of styles. Something that is often dismissed in manufactured pop music is the deep *love* of pop music that is displayed by the instigators and creative agents behind these records. I will reflect and comment on those agents, including myself, throughout this book. It is easy for people outside the creative field to criticize the process and the creators as 'faceless songwriting teams, who provide hits for bands such as Boyzone and Backstreet Boys at best ape, at worst parody' (Smith, 1999: The Guardian Newspaper). The aim of this book is to apply some of the existing systems of creativity, culture and cognitive psychology (Csikszentmihalyi, 1997; Sawyer, 2003; Bourdieu, 1996) to pop and BoyBands and to re-contextualize those empirically, illustrating some new academic commentary on the music industry, or what I refer to as a 'service industry'.

Was there a 'signature sound' (Zagorski-Thomas, 2014) for the BoyBands of the 1990s? Was there a signature sound for the Harding & Curnow (P&E) production team of the 1990s? I could say a simple 'yes' to both questions, but it requires a deeper examination to help define these in a schematic system such as the systems model for creativity (Csikszentmihalyi, 1997). Whilst approaches would be similar by different BoyBand producers, there were once again many different pop music styles and genres incorporated into what we might call the 'BoyBand sound of the 1990s'. If one can define the 1990s signature sound and turn it into a repeatable framework, can that framework be ethnographically updated for now, the 2020s and beyond? I will define and contextualize song arrangements for manufactured pop and BoyBands in Chapter 4; the music production methodologies for those genres are also mapped within the book (Chapter 5). The aim is to present an understanding of how the manufactured pop and BoyBand production process worked in the 1990s by providing an analysis of the creative process. As an industry practitioner in this field at the time, I am in a unique position to be able to interrogate and evaluate the theoretical frameworks in a way that has not yet been achieved.

Although British and European artists largely led the 1990s BoyBand phenomenon, it was kick-started by American BoyBand, New Kids On The Block (NKOTB). Renowned entertainment entrepreneur and music producer, Maurice Starr, formed NKOTB in 1984. The majority of their success was in the late 1980s through to 1994. As a reaction to the success of NKOTB in the UK, many music industry managers and entrepreneurs began to hold auditions to 'manufacture' BoyBands to rival NKOTB. The first of these was Take That, chosen and groomed by aspiring music industry manager Nigel Martin-Smith in 1989–1990. Shortly after that, music industry entrepreneur Tom Watkins signed East 17 to his Massive Management company, achieving a record deal for East 17 with London Records by 1992.

Upon leaving PWL Studios in early 1992, my production partner, Ian Curnow, and I were approached by A&R Director, Nick Raymonde, to remix and produce the last few tracks for the Take That debut album 'Take That & Party'. Tom Watkins also approached us to remix the debut single 'House of Love' for East 17 around the same time[1]. Towards the end of 1992, having accepted a management offer from Tom Watkins, Ian and I were given a clear choice by him to 'either carry on working with East 17 or work with Take That, but you can't do both' (Watkins, 1992, personal interview). The P&E decision to commit to Tom Watkins, Massive Management and East 17, was to set the destiny for Ian Curnow and I to spend the whole decade producing records for manufactured pop and BoyBands.

It is rare to find manufactured pop and BoyBand music production critiqued in studies by musicologists and researchers. Some have tackled the seemingly nebulous and elusive target for the subject of music production (Hepworth-Sawyer & Golding, 2012; Burgess, 1997; Massey, 2000). The fact that music production is a relatively new subject to academia explains the lack of study references but research data for students is being collected with the introduction of field study (Frith & Zagorski Thomas, 2012) and a useful network of industry practitioners and academics contributing to journals and conference papers; ASARP (the Association for the Study of the Art of Record Production). Zagorski-Thomas (2014) states that the main problem with academic study of music is that it has not sufficiently addressed the ontological question of how recording changed music and how that change needs to be incorporated into study. The creative power of collaboration (Sawyer, 2007) is illustrated throughout the book and I have adapted some of Csikszentmihalyi's (1992, 2002) suggestions of *flow* in the workplace to develop my 'Service Model for (Pop Music) Creativity and Commerce' (see Chapter 8).

There has been some commentary about what kind of person gravitates towards this profession, what being a music producer really means and how to make a living out of music production (Hepworth-Sawyer & Golding, 2012; Harding, 2010; Perry, 2008). Fundamentally, there is no one type of person that spends a lifetime on music production but generally they will originate from being a recording engineer (Harding, 2010), or a musician/arranger (Visconti, 2007); occasionally creative business people may also find themselves in the production seat and describing the process (Cauty & Drummond, 1988). Pop music culture focusing on specific periods in time, such as the 1960s and the music managers' perspective (Napier-Bell, 2001) or the 1970s music producers' perspective (Visconti, 2007) are balanced by the academic view of the culture and conflict in the popular music industry (Negus, 1992) and the analysis of popular music (Moore, 2003). Bourdieu (1984 & 1996) describes an individual with cultural, symbolic and economic capital in a way that gives structure to my descriptions of music entrepreneurs in the pop music cultural fields of the 1980s and 1990s. I have used Pete Waterman and Tom Watkins as examples of people forming creative recording studio teams due to their experience, knowledge, status and money from previous success in the same field and domain. The artist perspective is the most commonly researched and written about, such as Lefcowitz (2013) commentating on what we might now call an early BoyBand, for example, The Monkees: 'It is difficult to disassociate yourself from the project that defined you' (Lefcowitz, 2013, p.267). The popular cultural cynicism of 1980s Thatcherite Britain from a music producer and artist point of view (Cauty & Drummond, 1988) is a far cry from my own memoirs of that decade (Harding, 2010). Many commentaries are significant documents of the time and reflect on how culture has affected music and how music comments on the sociology and culture happening around the artist (Byrne, 2012; Visconti, 2007; Frith, 1996). We now live in a music world dominated by an abundance of what Adorno (1970) might suggest is 'low art', with television programs such as 'X-Factor' and 'The Voice'. Eurovision has almost become a pop subculture or 'aesthetic discrimination' (Frith, 1996) that is largely ignored by the majority of music fans. David Byrne (2012) believes music needs to have texture and that was lacking in his band Talking Heads as a three piece until Jerry Harrison joined as keyboard player/second guitarist to add that texture with contrapuntal parts to compliment what Byrne was playing on guitar. Byrne (2012) also calls the current styles of superstar concerts 'Karaoke Spectaculars', where only the vocal is live, most of the music is pre-recorded and the stage musicians are just there to 'dress the set' (Byrne, 2012, p.133). Live performance music is ephemeral and lasts a very short time whereas recorded music is with us forever.

The daily pitfalls for record producers working with artists and technology in the recording studio have been well documented over the last two decades (Burgess, 1997; Massey, 2000, 2009; Harding, 2010; Visconti, 2007). The creative production and technology process is collaborative, involving commercial and technical judgments between the producer, artist and creative business people such as record company artist and repertoire (A&R) executives and artist managers (Burgess, 1997; Harding, 2010; Hepworth-Sawyer & Golding, 2012). Technology in music production has grown to the point where many believe 'music production has been taken out of the hands of the few and placed into the hands of the masses' (Oboh, 2012). Possibly the ultimate in music production technology research and study leads to a plethora of 'how to' books that will claim to present a friendly user-guide that informs beyond the supplied manuals with the current music production software (Collins, 2009, 2013; Cousin & Hepworth-Sawyer, 2014). Zagorski-Thomas (2014) talks about schematic mental representations that the music listener builds on the basis of sonic characteristics and I will comment on examples of those with the 'PWL' sound from the 1980s and the 'Harding and Curnow' sound from the 1990s throughout this book. Moylan (2015) states that sound quality and timbre are important elements towards understanding and crafting the mix and I present a 12-step program to explain my own framework towards a repeatable mix system for the pop genre that starts with the vocals and then works down through the arrangement to the drums

This book has been structured to describe how popular music culture from the 1950s to the 1980s led to the 1990s BoyBand phenomenon. Chapter 1 considers how the early entrepreneurial role models of the 1950s and 1960s, for example Larry Parnes, paved the way for pop impresarios, such as Pete Waterman and Tom Watkins, both of whom feature in my commentary throughout and are the ideal team-leader choices for my 'Service Model for (Pop Music) Creativity and Commerce' framework in Figure 8.1.

Chapter 2 investigates the business models and structures around pop music production and highlights how critical it is for current pop music producers and teams to be entrepreneurial with business negotiation skills and knowledge. Chapter 3 outlines the concept of turning creativity into commerce and takes a fresh look at the current concepts, systems and studies into creativity, which is a huge subject and how we can relate them to pop music production. Chapters 4, 5 and 6 look into the heart of pop music production in the chronological order that we approach them: Songwriting, production recording and mixing.

Chapters 7 and 8 conclude the book by looking at case studies of some of my pop music productions in detail, followed by bringing everything up-to-date with my conclusions and theories that have developed from this study. For those interested in the studio technology of the 1990s pop and BoyBand explosion, this can be found in the Appendix. A further development of my 'Service Model for (Pop

Music) Creativity and Commerce' can be found in these published proceedings papers:

1. A 'Service' Model of Commercial Pop Music Production at PWL in the 1980s in *Innovation in Music: Performance, Production, Technology, and Business*' Edited by Russ Hepworth-Sawyer, Jay Hodgson, Justin Paterson & Rob Toulson. Routledge (2019).
2. A 'Service' Model of Creativity in Commercial Pop Music at P&E Studios in the 1990s in *Journal on the Art of Record Production – Proceedings of the 12th Art of Record Production Conference, 2017.* KMH, Royal College of Music, Stockholm: Sweden (KMH/JARP, 2019).

METHODOLOGY

This creative and cultural research book features my 'lived experience' (Husserl, 1927), reflecting on my songwriting and production work within the manufactured pop and BoyBand field during the 1990s. The research has followed the qualitative research method and explores the shared culture of manufactured pop and BoyBands. It is supported with interview data from other practitioners (music industry managers, artists, singers, songwriters and producers) and their reflections of the same field and domain. In some cases, my interview respondents and I have reacted to and commented on the current pop music cultural field in context with the subject. The data gathered from these interviews and my own reflections, sourcing documentation such as diaries, files, invoices and studio workbooks, has been analyzed using interpretative phenomenological analysis (IPA).

A PHILOSOPHY OF ART

The choice of technology can influence artistic and creative outcomes that then become successful and repeatable commerce. The subject of manufactured pop music is often criticized and mocked in the media and music industries as a quick and 'cheap shot' at making money. Many creative people will say that they did not enter the music industry just to make money, they entered it to be creative. Few understand the creative passion, the *love* of pop music and the motivation that practitioners require to achieve success in this field of music. This book questions the legitimacy and authenticity of these misconceptions and will explore the attitude and modus operandi to work in this much-derided area of music production.

I will discuss the following:

1. **BoyBands and manufactured pop are ultimately a creative phenomena:** Even if they arise initially from a purely business process.
2. **The choice of technology influences creative outcomes:** This theme has echoed through the ages, from art through to music and my

commentary (Chapter 7) on how the evolving music technology in 1992 manifested the platform to create the first East 17 hit single 'House of Love', is a case in point.

3. **A 'Service Model for (Pop Music) Creativity and Commerce':** Can be tested in this genre of music when there is harmony in the field between all the agents and a person with novelty in the music cultural domain. That person needs to be capable of divergent thinking and convergent thinking at the same time, possessing the ability to tell a good idea from a bad one. Then the settings are in place for a creative process that acquires 'all the dimensions of *flow*' (Csikszentmihalyi, 1997).

4. **'Group Creativity and Human Interaction':** This is discussed in Chapter 8 and will provide evidence of unexpected outcomes when testing a proposed formula working specifically towards the manufactured pop and BoyBand cultural domain.

INTERVIEW QUESTIONS

My interview strategy focused on a series of ethnographic questions proposed to my respondents. Participant observation outside my framework of questions has also been collected and analyzed using NVivo software and its coding (node) system. The questions were designed to keep the respondents focused with the subject matter and to draw out as much side material as possible, encouraging self-reflection from my respondents of their own memories of the period. However, I found that many people had not given much thought to this subject recently but once they were re-engaged by these and other questions that developed throughout the interview useful information began to emerge. Conducting a relatively small number of interviews has allowed me to analyze the interview transcription data from each respondent individually, which is a useful IPA method.

Table I.1 highlights the main themes of pop music culture and technology that emerged from the analysis of the seven one-hour (approximately) interviews with the participants. The interviews used a semi-structured auto-ethnographic approach and took place between 2014 and 2016.

Table I.1 Themes Derived from Interview Analysis.

Pop Music Culture	Technology and Process
Song formula framework	Music technology advances (1990s)
Producer roles and workflow	Studio spaces and equipment
Collective creativity	Production methods and work modes
Industry attitudes	Pop vocal recording
Social, economic and cultural capital	Signature sounds
Tacit knowledge	The sonic picture
Phenomena and cultural trends	

What follows is a list of the main questions posed to the interview participants that prompted the themes in Table I.1 and the in-depth reflections and discussions throughout the study.

1. **Is there a formula to create a successful manufactured BoyBand?**

     Is there a music composition and production formula for a BoyBand? This idea is rooted in the transcultural context of the 1990s and it is a useful question for musicologists, entrepreneurs, composers and producers to consider. The formula discussion has raised some interesting comments and further questions from my interview respondents around the compositional techniques and the music production technology used today in professional studios and home recording facilities, such as: What interactive media do composers and musicians in regional and international contexts use for the collaboration process? Do composition and even recording sessions need to take place in the same room any longer?

2. **What causes an average short shelf life of a manufactured pop or BoyBand?**

     Very few bands in this field and domain remain together for a long period of time, two or three albums in the 1990s was considered a long run for a BoyBand such as East 17, often there was only one album (OTT and Deuce). There were some exceptions, such as Westlife and Take That, who did split up and have reformed in recent years with some variation in the line-up (Robbie Williams returned for one album and tour). The E! entertainment television channel recently broadcast 'The Wanted Life', where the cameras followed the British BoyBand, The Wanted, to Los Angeles and they are shown preparing for and recording their third album as well as performing some hastily arranged concerts. The cameras reveal a BoyBand in chaos, breaking up under the sustained pressure of remaining successful. I have collected some useful cultural data from this question that will go towards understanding why manufactured pop and BoyBands do not seem to last.

3. **What impact does technology have on the artistic processes?**

     Only a few technology-minded respondents have been able to comment on this but it has been useful for the technology sections in this book. The technology-focused appendix has been compiled from responses to this question and will provide a useful resource for future researchers of 1990s music technology.

4. **What does it take to be a principal agent or actor in the phenomenon of a successful BoyBand?**

     Again, this question has been useful in allowing respondents to talk about the 1990s pop and BoyBand phenomenon in wider terms than the previous questions and it has allowed me to ask the respondents what they would do in today's pop and BoyBand marketplace.

CASE STUDIES AND ANALYSIS

Smith, Flowers & Larkin (2009) state that 'usually the most exhilarating part of the analysis is that which is completely unexpected' (Smith, Flowers & Larkin, 2009, p.113) and that has certainly been the case for me during workshops experimenting with my 'Service Model For (Pop Music) Creativity and Commerce' theory at The Westerdals University, Oslo (in 2015 and 2016). Once human interaction began during the group songwriting and production sessions, the results took a surprising turn.

INTERVIEW RESPONDENTS AND TIMINGS

Tom Watkins: East 17, Deuce and P&E manager, **1 hour, 11 minutes**

Ian Curnow: Producer, songwriter, musician and P&E partner, **44 minutes**

Matthew Lindsay: Music journalist and Tom Watkins co-biographer, **1 hour, 8 minutes**

Nicky Graham and Deni Lew: Pop producers and songwriters, **1 hour, 31 minutes**

Tee Green: East 17 session backing vocalist and vocal trainer, **1 hour, 29 minutes**

John McLaughlin: Pop songwriter, producer and manager, **45 minutes**

NOTE

1. The commentary for this project can be found in Chapter 7.

1

BoyBands and Pop Music Culture

1.1 INTRODUCTION: BUILDING TOWARDS THE 1990S

Pop culture is a broad topic that can take in many aspects from the creative industries. The term *pop art* came into common use sociologically during the 1960s when Andy Warhol explored the relationship between artistic expression and celebrity *culture* from his 'Factory' studio headquarters in New York. Sociologist, Howard S. Becker, was a forerunner to Warhol on the theme 'sociology of deviance and art' but it is the legacy of Warhol that influenced the pop music culture from the 1970s onwards. Warhol's pop culture consisted of collaborations in art, video and music. When I arrived in New York in the late 1970s his influence on the pop culture scene, including music, was still in evidence. I will use an academic framework around Pierre Bourdieu's arguments in *Distinction* (1984) and commentary on the value problem (high and low culture) in cultural studies by academics, such as Simon Frith (1996), to uncover how social constructionism created the BoyBand phenomenon in the 1990s. In terms of high and low pop music culture, my subject falls into the latter category yet still generates its own cultural capital in mass media creativity and commerce. This chapter will discuss pop culture in terms of the music industry and will focus on manufactured pop music. Simon Frith (1996) states that the value judgment of popular culture 'has been quite neglected in academic cultural studies' (Frith, 1996, p.8). Twenty years on from writing, there is a wider academic access of popular music culture with many diverse publications, such as *The Art Of Record Production: An Introductory Reader for a New Academic Field* (Frith & Zagorski-Thomas, 2012), which discusses music production culture for academic study. On presenting the title of my PhD thesis to a number of colleagues whom I would consider to have 'cultural capital' (Bourdieu, 1984) in pop music, the name Larry Parnes repeatedly came into the conversation as one of the first entrepreneurial pop music managers to manufacture his acts. Music journalist, Matthew Lindsay,[1] offered these views of the manipulative style of music artist management that led to what I call the manufactured pop and BoyBands in the 1990s:

> Going back to someone like Larry Parnes, a manager in the 1950s through to the late 1960s and a renowned gay Svengali, he had a roster of talent that he *groomed*. He changed their names, he styled them and they became something exotic; acts like Tommy Steele, Marty Wilde and Billy Fury. Then you had The Beatles; their manager Brian Epstein groomed them in their early days, he put them in Pierre Cardin suits etc and that made them a little bit more acceptable to the general public, a bit more mainstream as opposed to staying with the all-leather look that they had themselves. It's interesting to look back on The Beatles because they developed into something so singular as themselves that you tend to forget about that first phase where he took a rough slab of marble and chiseled it into something that was going to be palatable to everybody. Then you go into the 1970s and you've got someone like Tam Paton [manager of Bay City Rollers] who was doing the same type of grooming with them. Then you look at The Monkees and they were manufactured as a pop group in the USA to rival The Beatles. But what really constitutes a manufactured BoyBand? You could even go to The Sex Pistols and their manager Malcolm McLaren; to some people they were a BoyBand, and their image was totally outrageous, totally staged by McLaren.
>
> (Lindsay, 2014, personal interview)

Lindsay raises the question 'what constitutes a manufactured BoyBand?' and it is his view that the manufactured pop music culture started with 1950s British impresario, Larry Parnes, and could be applied to The Sex Pistols in the late 1970s. He suggests that they were also manufactured and groomed by Malcolm McLaren and could even be termed as a BoyBand by today's understanding of that cultural label. I am going to argue that this style of management for the manufactured pop music genre is both prevalent and necessary.

In his book about pop music culture, *The Long-Player Goodbye* (2008), author Travis Elborough describes Parnes:

> Larry Parnes, British pop's original Svengali, dubbed 'Mr Parnes, Shillings and Pence' by Fleet Street, prided himself on his ability to turn out boy-next-door stars that British teenagers, evidently a nervy breed back then, could identify with.
>
> (Elborough, 2008, p.176)

Appealing to British teenagers, especially young female teenagers and, importantly, pre-teens, has been the core market for BoyBands throughout the recent decades. Lindsay goes on to say:

> There's a good point to be made here about pop music because it used to be the underdog, it was something that was sneered at [in the 1960s and 1970s] which gave it a slight power that it didn't quite have. David Hepworth of the magazine *Smash Hits* [1970s–1990s] said that the pop wars have been won [in the 1980s] and we now live in a pop world and pop has now got the hard task of learning how to rule. In the 1990s was when it changed – there's something about pop music that gets the better of you, it might be trite, it might sound really conveyer belt production-wise and yet there's something about it that you find irresistible and I think that we got to the point in the 1990s where it got very bland.
>
> (Lindsay, 2014, personal interview)

Lindsay suggests that pop music was the 'underdog' during the 1960s and 1970s, not only in the pop culture media but also, I would suggest, in academic commentary. From a position of working with SAW at PWL Studios and being at the forefront of pop music 'ruling' in the 1980s, I would suggest that the 'bland' observation could find its origins within that decade and that the 1990s was a natural extension of that. I am not sure I agree that pop music became bland in the 1990s, I would say it was worse in the 1980s and that the 1990s dragged us out of that, especially with the Britpop scene, with acts such as Oasis and Blur. Lindsay's comment about 'conveyer belt production' refers to the PWL 'Hit Factory' studio in the 1980s. In his book, *The Art of Record Production* (2013), Richard James Burgess refers to pop producers as 'Auteurs':

> Auteur producers write the songs, play instrumental parts, lay down guide vocals, engineer, edit and perhaps even mix as well.
>
> (Burgess, 2013, p.10)

The idea of auteur emerges from film-making in France and was coined by American film critic, Andrew Sarris, in the 1960s to indicate a director who is viewed as the main creative force and who has complete creative control over the elements of production. Burgess suggests that this type of music producer and practice started in the 1960s at Tamla Motown, with teams such as Holland, Dozier and Holland (who wrote and produced for The Four Tops, for example) and continued with producers such as Gamble and Huff (The O'Jays, The Jacksons and Teddy Pendergrass), L.A. and Babyface (Shalamar, Bobby Brown and Paula Abdul) and Jam and Lewis (SOS Band, Janet Jackson and Alexander O'Neal). Today, those auteur characteristics can be identified in producers such as Timbaland (Justin

Timberlake and Rihanna), Kanye West (Jay-Z and Alicia Keys), Max Martin (Britney Spears and Katy Perry) and the collaborators they surround themselves with. My commentary on manufactured pop and BoyBands is focused on the UK and Europe as this has always been the main cultural and social environment where I have had the most experience and economic capital. Lindsay has further commentary to add about British pop culture from the 1980s that will help to lead us into the 1990s with comparisons to today's pop market:

> If you compare a 1980s Bananarama record to a more recent Saturdays record, you find that Bananarama made great pop records and The Saturdays just sounded too establishment [predictable and bland]. If you think about what happened with the explosion of pop in the early 1980s you had these artistic pop records that were huge; Human League, Duran Duran and Soft Cell. Pop music then was really edgy for a brief flurry of activity in 1981–82 and then along came PWL [1984–1990]. Then for a few years from 1990–93 it wasn't cool to be a pop star [hence the demise of PWL] – then suddenly you get Take That, East 17 and you get Britpop [Blur and Oasis]. Neil Tennant of Pet Shop Boys would say; remember their song 'How Can You Be Taken Seriously?' – that was about watching Bros on the Terry Wogan TV show and they were saying 'yeah, we're about longevity'. I think with pop music – things happen initially for an act and they are explosive but it's completely disposable. The brilliant thing about the Pet Shop Boys was that they didn't seem to be trying to make you like them. In the 1980s, in their imperial days, it was like: This is what we are, we understand pop music, we're incredibly literate in it [Neil Tennant worked for *Smash Hits* magazine in the early 1980s] and we don't have to make you like it.
>
> (Lindsay, 2014, personal interview)

This is one viewpoint that provides an insight into popular music values from somebody who has clearly been observing and commentating on pop music for a long time, in particular from the 1980s through to today. Lindsay has consciously followed Csikszentmihalyi's (1992) *flow* mantra of immersing oneself in the culture that you wish to make your creative profession, internalizing the subject to the point that you become an expert who is in demand by the industry. This is something that most successful music producers have done throughout their careers and we can become unaware that we are subconsciously judging music with an industry practitioner's ear all of the time. Another respondent with a valuable viewpoint of pop music culture was successful UK songwriter, John McLaughlin:

John: I think kids need pop bands; it's a basic need. It's what got me into music. I was 7 or 8 when my dad used to play the records on the radio and you see them on television. Now we've got One Direction and that's got kids into music and I believe you need that. For me it was pop and then punk and I think a lot more kids picked up guitars and stuff because of that.

Phil: Others have said to me that at the time, bands like Bay City Rollers in the 1970s, were not called BoyBands, it is only since the 1990s with bands like New Kids On The Block (NKOTB), Take That and East 17 that we started using the word BoyBands. Do you agree with that?

John: Yeah exactly, at the time they were just pop bands or teen pop bands but ultimately we can now say that Bay City Rollers were the biggest BoyBand in the world.

(McLaughlin, 2015, personal interview)

McLaughlin confirms my point that the term BoyBand entered pop music culture in the 1990s simply because of the sheer numbers of BoyBands that record company executives wanted to sign, and were therefore manufactured and groomed by music managers and entrepreneurs to meet the demand. This point is excellently illustrated in the Channel 4 television series 'Boyz Unlimited' (2000) where the A&R executive is seen to be in competition with other colleagues saying 'well he's got two BoyBands signed and I've only got one, so I've got room for one more'. I agree with McLaughlin that 'kids need pop' but they will desert them once they can no longer identify with them, hence we see a continuing 'factory line' mentality in creating 'the next big thing'. Longevity for pop acts now is less likely than it was in the 1990s as it is no longer part of the manufactured pop formula. Much of my 1990s pop and BoyBand production and songwriting successes were achieved under the guidance of creative music manager, Tom Watkins. Lindsay is currently researching Tom's career with a view to completing his biography and has these observations to offer on Tom's style of music management and attitude towards pop culture and the acts whose careers he guided:

> What I think is really important about Malcolm McLaren and Frankie Goes To Hollywood (FGTH) is what Tom Watkins did with Bros, by comparison, where he took something that was squeaky clean and added some edginess and the frisson of outrage; and you could say that was something of a hangover from what Paul Morley [music journalist and ZTT Records A&R] and Trevor Horn [music producer and ZTT Records owner] were doing with FGTH. Just adding the feeling that it's all a little edgier than what it really is. Even with 'When Will I Be Famous' Tom had to cajole Bros into singing that; I don't think they wanted to do it. It was a bit Hi-Nrg for them. But then they want to screw with the formula too quickly don't they? When Bros put 'Push' [their debut album] out and it's huge and they start saying 'no we

wanna be taken seriously – let's change direction, we wanna be U2'. Well the obvious answer is you're not! You don't have that credibility or sound or fans. So Bros made a second album that wasn't really pop music; it alienated all the fans. With East 17 there was an element of rogue anarchy going on, like putting the dog [the East 17 logo on their early releases] signs and posters up around London and getting ex-convicts doing graffiti all around London and then putting the dog sign on the front of the record sleeve [to connect the two together] with no name, created an air of mischief and a little bit of iconoclasm that was a hangover from punk or something like punk; it's mischief making. You don't get that with One Direction. It's almost like they've gone further back to the days of Larry Parnes where everything is sweet and clean – cookie color.

(Lindsay, 2014, personal interview)

When a band such as Bros behaves the way Lindsay is describing, it alienates everyone around them who have helped to achieve that initial success with hit singles and a successful debut album. We will see as this book develops that an identical story unfolds time and again with the manufactured pop and BoyBands I worked with in the 1990s. Tom Watkins' cultural capital in pop music throughout the 1980s was a perfect platform for him to make what was essentially a comeback with East 17 in the 1990s. His company, Massive Management, had grown with previous business partners in the 1980s to a relatively large workforce in their central London offices, running the affairs of the Pet Shop Boys and Bros. When Ian Curnow and I met him in 1992 his central London office had been closed and Massive Management was run from his home address in Maida Vale with Richard 'Biff' Stannard, his new young partner and aspiring pop songwriter and producer. The important thing for us about Tom was his tremendous enthusiasm and the industry network around him:

> Maurice Oberstein's [former CBS/Sony Music UK Chairman] quote was 'you don't need the brain of a lawyer, or the brain of an accountant or any kind of established administrative figure, you just have to have an understanding of it and have the ability to appoint the best people to do it'. You need controversy, you need Sigue Sigue Sputnik, Frankie Goes To Hollywood, you need an element like when Bros came up and there was a perfect pair of twins and Ken the idiot who was the most successful one of them all.

(Watkins, 2014, personal interview)

Throughout the 1980s and 1990s Watkins surrounded himself with major 'players' in the higher echelons of the music industry, such as Maurice Oberstein (CBS/Sony Records) and Lucian Grange (Universal Records).

Success in pop music terms could be measured in a number of ways, usually the UK charts for a UK act, but one important factor in the 1980s was success in the USA market. Nicky Graham, the songwriter and producer behind the Bros success, noted:

> I remember when we tried to break Bros in America and we played Madison Square Gardens supporting Debbie Gibson and backstage half the Epic Records team turned up with NKOTB and one of the reasons Bros never broke USA big was because the label resented paying the money back to the UK when they said they could create their own BoyBand i.e. NKOTB.
>
> (Graham, 2015, personal interview)

That gives an indication of the internal politics that existed with major record labels such as Epic/Sony Records during the 1980s and this continued throughout the 1990s. My own experiences of pop culture throughout the 1980s was somewhat colored by my time as chief engineer at PWL Studios from 1984 to 1992. Much of that time was spent participating in and observing the rise and fall of the Stock Aitken Waterman songwriting and production team. I had the opportunity whilst based at PWL to form a remixing and production partnership with Ian Curnow, an established session musician and music programer who had toured the world as MD of successful pop act, Talk Talk, throughout the early 1980s. My published book *PWL from the Factory Floor* (Harding, 2010) commentates on that 1980s period in detail.

1.2 1990S POP MUSIC CULTURE: THE RISE OF THE BOYBAND PHENOMENON

To the pleasant surprise of Ian Curnow and myself, our PWL departure made front-page news in the trade magazine, *Music Week*, during February 1992. In March 1992, the USA magazine *DMR* (*Dance Music Report*) wrote about our departure from PWL and announced P&E Music Ltd as our new trading company. The Appendix will describe the technology of our 1990s studio but the final piece required at that time to complete the P&E Music jigsaw, to empower our negotiating position and representation, was a new manager. That became Tom Watkins of Massive Management. Ian and I liked Tom's attitude towards us throughout the 1992 remix project with his act East 17's first single 'House of Love'; this project is described in detail in Chapter 7. One sign of a creatively talented A&R representative, manager or music entrepreneur, is that someone can drive a session or mix to a conclusion and have the 100% confidence to say to the room 'that's a wrap' or 'that's it' or 'cut' as an auteur film director would. I have only ever worked with three people that not only had the confidence to say that to me, or a room full of people at the end of a mix, but would also be proven correct by those tracks going on to be success-ful chart records. They are Pete Waterman, Tom Watkins and Tracy Bennett. Close seconds might be Louis Walsh and Simon Cowell but they never seemed quite certain or displayed the confidence of the other three.

It is a typical music industry scenario that when a pop music man-ager has one successful act regularly entering the music charts, mag-azines, television and touring, a well-oiled machine will develop around that act. A driver, security people, musicians, vocal trainers, management office back-up serving their every need, 24/7. All of that, if it is put into place efficiently, will allow the manager to spend time on developing a new act whether that is found or manufactured. East 17 were the first act manufactured and constructed by Tom Watkins during 1991–1992, signed to London Records and already charting by late 1992. By late 1993 the next Massive Management act was 2wo Third3. Signed to Epic Records by A&R director Rob Stringer, Tom Watkins had persuaded Rob and Epic that success was guaranteed with two boys on keyboards and a good looking singer, colorful imag-ing, fantastically styled, an electronic, modern pop version of Kraft-werk with commercial but sometimes bland lyrics and melodies. We did achieve some minor UK chart single success with 2wo Third3 but the project did not fulfill it's potential and their album remains unre-leased in the UK. It did not help that Tom's personal and professional partner, Richard Stannard, was the main instigator of 2wo Third3, he formed the group, chose the musical direction and wrote most of the songs. When the personal relationship between Richard and Tom finished, the project fell apart.

Nevertheless, these were exciting times for me; I honestly never felt Ian Curnow and I would have achieved this position so soon after leaving PWL in 1992. I was confident that P&E Music could run as a studio and business after all our experiences at PWL in the 1980s and I was confident we could have some success with remixes and production hits, especially after some of our early 1990s work with Simon Cowell. I would never have predicted the run of success we had by 1995 or the depth of involvement with projects such as Deuce and East 17. Lindsay suggests the state of pop music by the mid-1990s had lost quality control compared to the 1980s and 1970s:

> I think that acts like Steps and The Spice Girls in the 1990s was like turning pop music into Butlins red coats[2]. Those acts were like light entertainment whereas Bananarama in the 1980s, if you watched them there was something very real about them by com-parison. All of this also on what you think pop music should be – is pop music just something that is designed to sell records and if it does that then it's doing its job well – or is pop music something that you think is really important and capable of being an art

form? At its best I think that pop music is the latter. Tom Watkins is always saying it's the music business – it's business! And I think 'you're right' and the industry bean counters think that way but there has to be something else in there that gives it a spark or something. Abba had that – compare Abba to Steps and you would have to conclude that Abba's records are beautifully constructed, there's a lot going on in those records but I'm not going to be putting on a Steps record in 20 years from now but I will be putting on Abba's 'The Winner Takes It All'.

(Lindsay, 2014, personal interview)

There are some interesting comparisons by Lindsay between ABBA and Steps and it remains to be seen if history agrees with his values. It is likely that the artists, songwriters and producers who created the Spice Girls and Steps records would disagree with him. Lindsay, one might argue, underestimates the artistry, passion, time and dedication that pop music teams devote to their records. In terms of the Tom Watkins point about blaming 'the business', I was involved in the songwriting and production process of a manufactured pop group campaign in the mid-1990s behind the Massive Management act Deuce. It is an example of the sustainability and disposability of the pop artists of the time and shows where the business power for these projects really lay. Tom Watkins meticulously planned the whole campaign[3], everything from image to styling, songs and musical direction. P&E spent a lot of time during 1994 songwriting and recording for the project in order to persuade Tracy Bennett at London Records to sign Deuce, this was sealed by the autumn of 1994. We carried on from that point to create a planned sequence of singles and a number of 'album filler' songs where we included the band on three song co-writes; 'Let's Call It A Day', 'I'll Be There For You' and 'Kiss It'. The long-term marketing strategy was: First single, 'Call It Love', in February 1995, to establish the group in the *Smash Hits* magazine and the Saturday morning television programs, then to achieve a good number of dance remixes to help establish Deuce in the pop and gay clubs in the UK so that Tom could arrange for the band to tour those dance clubs to promote the single. All of this with the hope that the single would at least break into the Top 40 UK singles chart and help to pave the way for entry into the UK Final for Eurovision 1995. The second single, 'I Need You', was to be the real breakthrough single, hoping it would become the UK's representative song for the Eurovision song contest final in May 1995 and to break into the top 10 or 20 of the UK singles chart, leading on to releasing the album in the summer on the back of the Eurovision and UK success of the second single. By the autumn of 1995 we would release a third single, 'On The Bible', from the album to retain the momentum of sales of the album through to Christmas. The plan fell apart when Deuce failed to win the UK Eurovision entry. The second single 'I Need You' did reach #10 in the UK singles chart and 'On The Bible' achieved top 20.

It was therefore a shock to be told by Tom that London Records had decided to release Deuce from their contract and not continue on to the second album. The album had peaked at #18 in the UK album charts on 9 September 1995, Ian and I thought that was a reasonable performance. Clearly too much marketing money had been spent compared with the income from the hit singles and album and it was summed up concisely by music industry executive, Jon Webster, in his Music Week column on 2 December 1995:

> The Deuce album may have charted at #18 but even after three top 20 hit singles, it sold less than 25,000 copies in the UK and no significant impact abroad. After a label spend of £250,000 or more? The figures just don't add up for the next album. Clearly that was the view of the London Records accountants.
>
> (Webster, 1995)

Another Music Week quote in November 1995 was 'Deuce to leave London Records despite hat-trick of hits'. London Records terminated the deal after one year by mutual agreement with Massive Management; Colin Bell of London Records said that the deal was too expensive to carry on as the outlay did not match the results. Tom said to us that he felt the image of Deuce (very pop and sugary) did not fit the 'Joe Cool' image of London Records. These experiences inform the creative practice of a team like P&E Music and one has to learn to be responsive to the dynamics of pop music culture. There is the need after a project like Deuce to pick yourself up, dust yourself down, absorb the experience and move on to the next project. This is much easier for the songwriters and producers to do than the artists in the group.

After the pop explosion of the 1980s, dominated by the SAW/PWL sound and records, something had to change for pop music in the 1990s. PWL tried to copy the American swing-beat sound of the late 1980s but it did not work as well for them on the latter Kylie Minogue albums, such as 'Rhythm Of Love' and 'Let's Get To It'. Both albums failed to match the chart success of the earlier PWL produced Kylie albums and soon Kylie would leave PWL Records to sign for a record label with more credibility. The UK Acid House scene exploded in the early 1990s but this was too mature and club-based for pre-teens and teenagers too young to get into. Smash Hits magazine had come to dominate and inform pop music fans by the early 1990s and one of their favorite bands were NKOTB from the USA. Clearly the time was right for a UK version of NKOTB and Take That were manufactured by manager, Nigel Martin-Smith, to be exactly that. To conclude this pop music culture section of the 1990s I will turn again to songwriter John McLaughlin to support these ideas:

> I think in the 1990s the time was right for BoyBands for a few reasons. Coming off the back of the 1980s everything got a bit

serious and electro [electronic dance music] was dying out. Then I think, for young kids [looking back], there was a moment there to introduce to the market, shiny, nice BoyBands. So I think it's all about timing and I think then, in the 1990s, it was all about TV. All the TV channels were geared towards kids, so therefore you had an opportunity to put younger music on TV on Saturday morning and that's what helped break bands and then when you were recognized on TV and the video support, then you're in the papers and all those things add up to the perfect storm.

(McLaughlin, 2015, personal interview)

John describes the media culture around pop and BoyBands of the 1990s accurately and it is clear that BoyBands were 'of their time' in the 1990s. As I have indicated around the Deuce campaign, promoting acts with children's Saturday morning television shows and magazines such as *Smash Hits* was often enough to chart a single. Those outlets no longer exist.

1.3 BOYBAND CULTURE

I will look at the music composition and production frameworks to create a successful BoyBand in the 1990s, but first I will outline the ideal BoyBand members as this can affect the songwriting and production styles. This is an example of how a BoyBand might have been formed in the 1990s:

> One gay member/one blonde member/one songwriting member/one cute member and one hunky member. That was and still is an ideal five and hopefully at least two of the members would be able to sing and the other three members able to dance. An experienced and enthusiastic manager was essential and would probably have been the person choosing all of the personnel from auditions.

This formula could vary, especially if there were only four members; in addition the songwriter member was not vital. There was a fictional BoyBand featured in the UK Channel 4 television program 'Boyz Unlimited', a comedy series that broadcast in early 2000, written by Richard Osman with Matt Lucas and David Walliams after the BoyBand phenomenon had peaked. Ian Curnow and I supplied all of the music for the series. It shows that the genre was able to mock itself. East 17 manager, Tom Watkins, also likes to mock the genre in a jovial way and states:

> 'If it's going to appeal to gay men it's going to appeal to girls, I will always advocate that we were peddling sex'.
>
> (Watkins, 2014, personal interview)

One of the things people misunderstand about BoyBand culture is quite how hard their management and record companies work them. It is a 24/7 lifestyle that allows very little time to relax, refresh and recharge their energies. Once the 'work' finishes there is generally a family and friendship circle that also needs to be satisfied with information and exclusive updates on the artist's creativity. There will be more commentary in Chapter 2 about pop and BoyBand lifestyles. I wondered what the current situation is like around forming a BoyBand in 2014; is the favorite number still five as it was in the 1990s? Have the audition methods changed from what I experienced in the 1990s where the manager would choose the members by their looks and the producers would audition the vocal abilities over a pre-prepared backing track? One of my interview respondents, Ian Curnow, was fortunately in the middle of that process and brings us up to date with the current BoyBand culture:

Ian: Five is still considered the best number for BoyBands because it's uneven numbers and you've got enough for everybody. The best audition method for us [in 2014] is to get them in the room and ask them to sing something – no accompaniment – really naked, because if they can't stand in front of you in a room and do it, how are they going to do it in front of a crowd? We've had some that are so nervous they can't do it and some that are confident but clearly can't sing well. I remember we (P&E) turned down a BoyBand that had been put together in the 1990s by some guys that rented out cars and wanted to invest their money in a BoyBand project – I can't remember the band's name but they spent all the money on getting the boys tanned on a yacht and looking great but they hadn't got a record yet, so we turned them down. That's what all these current BoyBand managers are doing now. Back then you had managers such as Louis Walsh [Boyzone], Tom Watkins [East 17] and Nigel Martin-Smith [Take That] and they had the understanding of how to push the bands through but you couldn't do that now. So the difference between then and now is that we are trying to do it properly. Then [the 1990s] you would put a showcase on and invite the labels to come and see the band [once you had songs and styling done] and get it signed if you'd done a good job. Now you haven't got a hope in hell of getting signed unless you've got the twitter numbers, enough numbers and activity visible because that's what A&R people do   they get a sniff of something then they go online and do their research. If they think it looks like it's got legs they want to get in there and get it and so you need to create that sort of buzz in 2014 to get signed.

Phil: And for the formula does the look have to be different with five matching boys?

Ian: Not really but you need a spread – you can't have all dark haired models – it's more important to have five characters than it is to have five catwalk models. You can't have one that's five foot and one that's six foot two; they've got to look good as a group together. Part of the audition process for us is to narrow the audition down to eight or nine boys and then go out into the car park and take pictures of them in groups of five just to see how they look together and match up. You want a spread of blondes and brunettes and you don't need a gay boy but we do need to know if they are.

(Curnow, 2014, personal interview)

That gives an interesting perspective when comparing the BoyBand culture of the 1990s to now. Curnow states that 1990s entrepreneurial BoyBand managers, such as Tom Watkins and Nigel Martin-Smith, are lacking from the current scene and auditions taking place with just a cappella vocal renditions ring true to what we have seen on television shows such as X-Factor and Britain's Got Talent in their early days. I asked a similar question to British songwriter John McLaughlin, from the point of view of manufacturing a pop or BoyBand now compared with the 1990s:

Phil: And the formula now in terms of image, music etc?

John: I think it's always been there, when you look back to Phil Spector, The Ronettes, after that Bay City Rollers in the 1970s, even The Ramones have said that they wanted to be like Bay City Rollers, got the leather jackets and jeans – that's all they had. They took the chant ideas from The Rollers like 'H.A.P.P.Y.' and ripped off Bay City Rollers for America. They've said it in documentaries that they wanted to be like Bay City Rollers but turn it into punk rock. 'Jabba Jabba Hey' and 'Rock, Rock, Rock n Roll Radio' – trying to be like Bay City Rollers chants from earlier in the 1970s.

Phil: What was the reason for only having three members in 911 during the 1990s? Was that a conscious decision?

John: Yes, we thought at first five members but then we thought – hold on – that means another car and we were skint then so it worked with 1 car, 3 kids and a driver. The driver was Steve Gilmore [911's manager] and that still left room for me, a completely punk rock attitude to a BoyBand. We made the songs in a flat in Patrick in Glasgow's west end. We got the three boys together and they all stayed with Steve in his house and we were pressing up records, doing school tours in one car. It was a more independent punk rock attitude than you would ever imagine trying to get 'scores on the doors'. So 911 were more punk rock than a lot of punk rock bands, they made it with no money. That's why I love Malcolm McLaren. The Sex Pistols were a BoyBand you know, put together by Malcolm McLaren and Vivian Westwood, and they had an amazing affect. 'Never Mind The Bollocks' was put together by the genius of Malcolm McLaren and Vivian Westwood and I think they deserve a lot of credit for it. Malcolm McLaren saw The Ramones at CBGB's in 1976 and The Ramones were trying to be Bay City Rollers. So he formed a BoyBand to help sell his girlfriend's clothes and that's why the tartan thing was big with Vivian Westwood because all of it was based on Bay City Rollers.

                                        (McLaughlin, 2015, personal interview)

I do not think many people would spot the influence of 1970s Scottish pop band Bay City Rollers on American punk band The Ramones. McLaughlin is not my only respondent to talk about The Sex Pistols and their grooming by manager Malcolm McLaren. I did not expect to hear these comparisons and influences to the BoyBands of the 1990s when I started my interviews for this research. Here is Lindsay's view on the BoyBand 'label' and genre in the 1990s:

Phil: Why was there such a big explosion of BoyBands in the 1990s?

Matt: It depends what you see as being a BoyBand – take The Spice Girls, we all see them as a GirlBand but were they year zero for the start of GirlBands? Well, Bananarama were a GirlBand in the 1980s. With BoyBands in the 1990s I think the media came up with that name because it was the perception that they were being manufactured [therefore they had to have a derogatory tribal name]. You look at BoyBands in the 1990s like Boyzone and you immediately think manufactured. Something that's been very chiseled for mainstream success, it's almost like the members have been plucked and overtly put together by an external force whereas Bananarama, they were flat-mates and it was a lot more organic.

Major record labels seized control in the late 1980s and punk's influence on pop, an influence that had proved so vital to the pop explosion of the 1980s receded into the background. A manufactured BoyBand was a record label's wet dream, easily manipulated, pliable product at least in theory. A far cry from the independent spirits that had shaped pop music in the early half of the 1980s.

On a deeper note, I think AIDS may have something to do with it. When AIDS hit, pop seemed to become more conservative, less 'weird'. A BoyBand, squeaky clean, wholesome and 'healthy' existed in a universe entirely removed from these harsh realities, it was pure escapism. The same thing happened with the PWL sound, the way it shifted from Hi-Nrg gay underground to something a lot softer.

                                        (Lindsay, 2014, personal interview)

The grooming of pop acts began with Svengali manager Larry Parnes, as I have highlighted at the start of this chapter. It has clearly continued since the late 1950s throughout the decades, with acts such as The Monkees (1960s), The Sex Pistols (1970s) and all of the PWL acts (1980s) but I would emphasize that it was never more prevalent in pop music culture than in the 1990s and the acts that I was involved with, such as East 17, Deuce and Boyzone. Producers and songwriters, Nicky Graham and Deni Lew, have similar views to Lindsay and they discuss the 'BoyBand' nickname that the media directed at the 1990s bands such as East 17 and Take That. What were the origins of the name and how far back in pop culture can we go? My point is that the BoyBand name really only started in music culture conversation in the 1990s and since then people have labeled earlier pop bands containing boys by the same generic name:

Phil: Would you have called Bros a BoyBand?

Nicky: Yes as I call a band with boys in a BoyBand, whether they play instruments or not, but BoyBand is really just an alliteration for the genre.

Deni: But there were bands like The Jacksons and The Osmonds that I wouldn't call BoyBands – but I don't think the phrase BoyBand was used in the 1960s/1970s and 1980s, not until the 1990s did we hear it. Who coined the phrase? I think it is a British phrase, I don't remember Americans saying that. They used to say Boy Groups I think. Maybe Take That were the first to be called a BoyBand? I think it was the media that maybe coined the term BoyBand around the early 1990s and to me a BoyBand is meant to sing and dance so I don't understand how One Direction have got away with it for as long as they have because they don't dance – they can't dance! For me, growing up, NKOTB were amazing, Take That, Backstreet Boys – all amazing singers or at least three singers in each band.

Phil: It did seem that when the Americans created a BoyBand – like NKOTB or Backstreet Boys or NSYNC, they would break them successfully worldwide whereas Take That and East 17 from the UK never broke America – why was that?

Deni: Make no mistake, those USA BoyBands were better, those bands were more talented, always really good singers. Same with New Edition and Boyz II Men, they both had great singers that then became solo stars such as Bobby Brown.

Nicky: The pool of talent in all aspects was always much better in America, the pure talent itself, the choreography, the songwriting even though the songwriting latterly went to Max Martin and those types of people in Europe and Scandinavia.

(Graham & Lew, 2015, personal interview)

Nicky Graham and Deni Lew manufactured a BoyBand in the mid-1990s called Code Red, who were relatively unsuccessful in the UK even though they secured a good recording contract with Polydor Records. Nicky Graham gives us an insight as to why that success was not forthcoming:

> Every day they were here in Parsons Green rehearsing, then eventually we got them gigging, doing performances and eventually onto the all-important Smash Hits tours in the 1990s. The thing with Code Red was that because we couldn't get Polydor's money, because it was all going to Boyzone, we had to find an alternative market [to the UK]. So, by coincidence, somebody in Bangkok had gone into a record shop and asked for a record by Code Red and found a record that had been imported, it was the Eurovision entry that we had done for the UK finals that was released as the first single, 'I Gave You Everything', written with Deni Lew, Wayne [Hector] and myself. It was the year that Gina G represented the UK in the Eurovision final in 1996. So somebody at Polydor phoned us and said there seems to be a market for your band in Thailand, so will you go out and do some promotion there? Deni went to Thailand with the band for a couple of weeks and discovered that they loved our style of music. So all of a sudden Polydor went 'ah great we don't need to worry about breaking Code Red in the UK, we'll break them in the Far East'.
>
> (Graham, 2015, personal interview)

Graham's commentary on Code Red's rehearsal schedule indicates the time and dedication required when forming a BoyBand in the 1990s. Code Red's style of music was a little more Soul/R&B-orientated than the other British BoyBands of the time. I asked Tee Green, our East 17 session backing vocalist on the first two albums, about his job with East 17 on the road in the 1990s during their tours. He described his role as vocal coach, masseur, alarm clock and nutritionist. When I asked Tee if he would consider helping to form a new BoyBand now [2015] he answered very assertively:

> No because it's the most thankless task there is, I've seen it myself first hand, what an ungrateful set of bastards a BoyBand can be – most of them don't want to be there. They want to walk the bridge but they don't want to build it, so no I wouldn't do it, I'd do it for my daughter. If a team of business people came to me though with money and a good proposition where I'm in control then I would do it and I'd get Randy Hope Taylor on bass, a great drummer, Phil Harding to produce it, string arrangements by Ian Curnow and I would know that what I was about to build would be unstoppable. The problem is the music industry people who want to spend more on cocaine and cabs than they do on vocal coaches; the three 'c's.
>
> (Green, 2015, personal interview)

That says a tremendous amount about the negative culture that can surround a BoyBand and how those that have experienced it would want all of the required elements in place before embarking on the journey again. I call that 'having all the links in the chain working together'. Tony Mortimer of East 17 agrees with my earlier point about the difficulties of the 24/7 lifestyles for BoyBands and Tony has stated that Tom Watkins understood the importance of the music. Tony Mortimer has an analogy of describing the song as the 'cog of a wheel' and everything else in pop music culture spreads out from the spokes of that wheel; that is an interesting point of view that I have never considered before. That is his version of my 'links in the chain' analogy. In 1992, Ian Curnow and I had no idea how significant the BoyBand culture would be to us. Having achieved chart success with our East 17 mixes and productions from their first album 'Waltham-stow' in 1992, we thought it was time for Tom Watkins and the band to embark on a second album. To our surprise they decided to record a new single 'It's Alright' with us in January 1993 for further inclusion on the 'Walthamstow' album as it was taking time for Tony Mortimer to write enough songs towards the second album 'Steam'. Tom viewed it as a vital 'bridging' single between the first and second albums. 'Walthamstow' was considered finished in 1992 but to buy the band and label time to prepare for the second album it was decided that a killer hit single would be needed. The song starts with a ballad-style introduction and first verse; indicative of what was to come on future songs such as 'Stay Another Day'. After the introduction and down-beat verse, the vocal chant 'It's Alright' bursts in with heavy dance beats in support, reminiscent of the first single 'House of Love'. 'It's Alright' was a tremendous success and achieved #3 in the UK charts. Tom Watkins spent many hours in the studio with Ian and I on 'It's Alright' and pushed us hard to create the dynamics and excitement that he felt the track needed to be commercially successful. This was an early indication to me of the kind of control that Tom Watkins would want in our future projects together and to be honest, it was exactly the kind of leadership and inspiration that Ian and I needed to achieve our best results as producers. Tom Watkins admitted his controlling nature during my personal interview with him in 2014.

By July 1994 we were finally working on the East 17 album 'Steam' after an aborted and controversial start with Richard Stannard and Matt Rowe producing the tracks, 'Steam' and 'Around the World'. The fallout from the breakdown of the personal relationship between Tom Watkins and Richard Stannard earlier in 1994 meant that Ian and I were drafted in to remix the track 'Steam' and commissioned to carry on and produce more tracks for the album. After remixing the single 'Steam' we moved on to producing songs from their inception, using Tony Mortimer's song demos. From Tony's point of view, Tom would always feel it best if Tony went into demo studios with co-writers such as Dominic Hawken to lay down ideas for us to work on.

Our 'inherited' co-producer Rob Kean[4] also got the chance to join in on some of these writing sessions and hence is credited for co-writing some of the songs. My interview with Tony Mortimer in 2014 confirmed what Ian and I always suspected, that there was minimal input from Rob Kean on those writing sessions. Ian and I felt that we deserved to be given a chance to co-write with Tony and I remember a particularly fractious telephone conversation with Tom Watkins over this issue. During this phone call I had to very strongly put the case for Ian and myself to be given the opportunity to co-write with either Tony Mortimer or Brian Harvey. Tom said he thought that we were too busy with all our other productions and remix work to make time for songwriting. He also did not like the fact that we had signed our publishing to Paul Curran at BMG Music before we had signed our management to Tom and he would have preferred us to be signed to his own publishing company 'Porky Music' through Polygram Music. Somehow I managed to get Tom to concede on at least two songs on this second East 17 album and these were 'Let it Rain' and 'Generation XTC'. 'Let it Rain' would become the third single taken from the album and a top 10 UK chart hit; this delighted our publishers BMG Music. All of this commentary gives an indication of the politics and democracy behind the scenes surrounding pop and BoyBand culture. Lindsay comments further on his views of Tom Watkins and the way in which Tom handled BoyBand culture and success in the 1990s:

> Going back to Tom Watkins, I don't think his genius lies in his musical ears; it's more about the packaging, the design and his entrepreneurial skills. The problem with the kind of control that Tom had over projects like East 17/Deuce and 2wo Third3 is that you can manipulate everything in the early days and you can get what you want as an entrepreneurial manager but then you can get bored with that very quickly and that never happened between Tom and Pet Shop Boys but it did happen with Bros and East 17 and this could also be one of the reasons for the short shelf-lives of those manufactured acts. East 17 are a classic example of just not knowing where your strengths lie and that's what really tore them apart and that's why they've never had that resurgence in popularity that Take That have. Take That know where their strengths are and they've played to it but East 17 never did.
>
> (Lindsay, 2014, personal interview)

An overview from Lindsay on how he felt the style of manipulative management displayed by Tom Watkins would work for some acts (East 17 and Deuce) but not for others (Pet Shop Boys). I highly valued the level of musical input from Watkins on our projects and would give it more credence than Lindsay indicates. Watkins spent many hours in the P&E Music studio, as described with East 17's 'It's Alright' single, steering and guiding us to construct the musical direction he envisaged for the records we were producing. Here are some further comments on the 1990s BoyBand culture from managers Tom Watkins and Louis Walsh:

> **Phil:** For me what started the BoyBand phenomenon in the 1990s was NKOTB.
>
> **Tom:** Ah, well, yes, because of the dance routines. The philosophy was very simple; OK, we're peddling sex, we have got a very good perfume, we have got a very good box that the perfume goes in, we are very nicely positioned on the shelves in the shops that are selling our exclusive perfume OK and it has got desirability about it. That's all I ever said. You can't take one of the components without endorsing it or seconding it with other equally as talented areas. It all started with the music. I have often been quoted as saying it is a 'necessary evil' right, because I knew how important it was once you had those quality goods.
>
> **Phil:** How did you have the hours in the day, seven days a week to achieve so much?
>
> **Tom:** Lots of cocaine! It is very simple, if you are going to be very successful, that's why I swore that I would get out at 50, I knew I was getting very close to getting worn out because it was a concise effort. You give up your social life, apart from the odd fuck that you might buy, steal or whatever, there is no private life and there is no time for any of that stuff.
>
> **Phil:** The media coverage of the battle between East 17 and Take That, similar to the Blur / Oasis media battle, do you have anything to say about that?
>
> **Tom:** We did it with Jules Smith at *Top Of The Pops* magazine and Alex Cadiz at *Smash Hits* magazine. I definitely did manipulate it, when we used to see the boys on the road [East 17 meeting Take That] everyone was laughing and joking.
>
> (Watkins, 2014, personal interview)

This is honest information from Tom Watkins on how he would retain such a busy schedule, which he enforced upon himself by demanding so much control over everything around his artists. I had suspected his manipulation of the media over the public rivalry between Take That and East 17. One of the remarkable things that happened in the latter part of 1995 was that once the news of our split with Tom Watkins became public, we had a phone call from music manager Louis Walsh asking us to produce Boyzone. Louis had already been achieving success with Boyzone since early 1995 with the single 'Love Me For A Reason', a typical BoyBand cover sound of the Osmonds 1970s hit. Louis Walsh (2007) notes in his book *Fast Track To Fame*:

> I'll never forget receiving the mid-week charts that Tuesday morning and discovering they [Boyzone] were #2. East 17's huge hit

'Stay Another Day' kept us from going to #1. But it didn't matter because being in the top 5 with our first single was beyond any of our wildest dreams. 'Love Me For A Reason' went on the sell over 700,000 copies.

(Walsh, 2007, p.39)

This shows the importance of the charts as a measure of success to the music industry. Pop music culture is a dynamic and unpredictable force and quite simply, we would use the charts as an indicator of what worked and what did not. The record companies and the managers used that judgment factor and it reflects the producers' difficulty in gaining a position of trust and the ability to interpret the demands of managers such as Tom Watkins and Louis Walsh. If a #1 record was achieved, as Walsh is indicating, it breeds a desire within those industry observers to work with the team that achieved that success.

1.4 GAY CULTURE: IS THERE A GAY AESTHETIC TO POP MUSIC?

Gay culture and pop culture have been linked as far back as the 1950s as we have heard from my interview respondents regarding manipulative music managers such as Larry Parnes through to Tom Paton, manager of Bay City Rollers in the 1970s. On the BBC Four television program 'The Music Moguls' (broadcast 15 January 2016) 1960s pop impresario Simon Napier-Bell states that for the first five years of the 1960s gay managers, such as Larry Parnes and Brian Epstein, were well known and that most of the successful pop managers at that time were gay. They were already leading double lives because homosexuality was illegal at that time and had to be hidden. This put them in a good position to be entrepreneurial pop managers, dealing with the artists (mainly young boys) on the one hand and also dealing with the record company executives who were from the same public school backgrounds and middle class as the managers. Therefore these managers were a good bridge between those two. In the 1980s Pete Waterman would target the early SAW pop productions at the gay clubs of that era, knowing that if that market could be won over then it was a solid base to build on to enter the pop dance clubs, followed by pop chart success. Early examples of that are the drag queen Divine's 'You Think You're A Man', which was successful in the gay dance clubs and peaked at #16 in the UK pop charts in 1984. Following that was the Dead or Alive single 'You Spin Me Round' that took a similar route through the gay dance club scene and peaked at #1 in the UK pop charts in March 1985. There is full commentary on this type of gay-to-pop crossover PWL/SAW success and methodology in my book *PWL From The Factory Floor*

(Harding, 2010). Music journalist Matthew Lindsay also has much to say, as an observer of this subject:

> The gay scene is like the litmus test for BoyBands and Take That played to that and were massive because of that start point. Pete Waterman has talked about the relationship between gay men and the young teenage girl fans of pop acts and BoyBands, saying that the men can't bear them [the girls]. Is that something that's been shaped by PWL? I think that gay culture was really changed by PWL in the 1980s; the sound of gay clubs became synonymous with the PWL records. If you look at the music in gay clubs before PWL it was a lot harder, a lot more underground, and a lot edgier and then PWL came along and it made the music in gay clubs very bright and cheerful. Obviously it came from Hi-Nrg music, but it got an extra dose of pop.
>
> (Lindsay, 2014, personal interview)

That was Pete Waterman's philosophy: Take something that is underground, turn it into 'overground' and propel it into the charts by adding more pop sensibility to the records. We took Chicago House music at PWL and turned it into 'London House' with Mel and Kim's 'Showing Out' and 'Respectable'. This highlights Waterman's ability to anticipate the dynamic and seismic shifts of pop music culture of the time and his early entry into making records specifically for the gay clubs of the early 1980s brought rich rewards for the SAW production team. Moving into the 1990s, entrepreneurial manager Tom Watkins was not shy of such blatant manipulation of the gay dance club scene in the UK as a crossover to pop chart success:

> The culture side can be simply summed up in one word – it's homosexuality right? It's basically, I don't know whom the quote is from, that 'you can't go and buy slices of penis but you can buy slices of vinyl which are called records'. And I think what you must remember with the BoyBand culture is not only does it appeal to little girls but because of gay men they are only exposed to one kind of music basically, and whether they are fifteen or fifty five, they still go to clubs like Bang and Heaven [dance clubs in London] because it's sexual. Selling pop music has always been sexual whether buying a Kylie Minogue – good god she's beautiful and a diva and all the rest of it and straight men like her and gay men like her as an iconic vehicle. But it was born out of a freedom and expression around Section 28 and what was happening politically. I think gay men were given a freedom. We ran a service called 'Ram Male' at Massive management which was basically a hyping service, no one would remember much about that except me but what we used to do was poll every gay club in

the country and we used to get every DJ and person who ran the club to go and buy ten or twenty records in their towns, fill in a return and we would pay them £1 each so we could always hype those records. Well the records that we were involved with were BoyBands.

(Watkins, 2014, personal interview)

It was clear to me throughout our association with Tom Watkins from late 1992 through to October 1995 that he was part of a music industry 'gay mafia'. He liked to say there was a demographic of music fans that would buy records by selling them slices of penises and the 1990s was a period that represented a new thinking and liberation on the gay scene; 'it was OK to like divas, it was OK to like Kylie' (Watkins, 2014, personal interview). Here is how East 17 struck Tom in the early days of them coming together. East 17's songwriter Tony Mortimer had been playing Tom his song demos for some time and Tom's suggestion was to surround himself with some like-minded friends who could sing and dance to form a BoyBand:

It was rough trade. I looked at them [East 17] and I knew they were sexy. I knew that these were the kind of boys that the girls would shit themselves for. Most of the queens I was hanging out with in clubs would die for them. Tony Mortimer took off his shirt and those photos we took at Harrogate Baths, with Tony and them virtually naked, I think are probably on porn channels even now. I tested it so much to see how far you could go because if it is going to appeal to gay men, it is going to appeal to girls. The girls still believe that half of these fucking people are straight. Every BoyBand [of the 1990s], every one of them, had a gay guy in it.

(Watkins, 2014, personal interview)

Compared with Take That, East 17 were not heavily marketed to the gay scene as the band were not comfortable with performing at gay clubs or appearing in any way to appeal to gay men. It did not suit their London East End 'rough-boy' image that was constructed to be a contrast to Take That. The bands would become rivals very quickly after each had some early hit records and both would dismiss reports of a gay following. Lindsay compares Take That in their early days to male striptease act The Chippendales:

Take That were massive on the 1990s gay scene and their first video for 'Do What You Like', it was gay porn, they were rolling around having jelly smeared on their bums and it was interesting that they were marketed that way early on in their career. You can't imagine East 17 ever agreeing to do that sort of thing.

(Lindsay, 2014, personal interview)

Some succinct and honest commentary there from Watkins and Lindsay on the links between gay culture and BoyBands. I asked producers Nicky Graham and Deni Lew if they felt there needed to be a gay
member in a 1990s BoyBand:

> **Nicky:** Yes – there always is.
> **Deni:** I don't think there has to be but there has to be one boy
> where gay men think he is gay [to appeal to them] or to be able to
> identify with him, definitely.
> **Nicky:** The little boy in Boyzone [Stephen Gately] – he was the
> one they all loved and he happened to be gay.
> **Deni:** He was gay but I don't think he has to be gay – but gay
> men have to think he is or could be.
>
> (Graham & Lew, 2015, personal interview)

The indication from Lew is that the perception of a BoyBand member being gay is more important than whether he is or not, and as
Ian Curnow commented earlier, the management need to know if a
potential BoyBand member is gay so that it can be managed in the
media. There have been multiple gay currents in music, from disco
music to today's ubiquitous Electronic Dance Music (EDM) but it
may be wrong to identify a gay aesthetic exclusively to BoyBand culture and music just because there are so many associations asserted
by my respondents such as Lindsay and Watkins. Other gay-pop genres, such as 'Glam Piano', have been examined by Geoffrey Himes
(Himes, 2015) and have been traced back to the New Orleans bars
of the 1950s. He also points to the lesbian-folk-song movement self-
described as 'Womyn's Music'. Certainly, we can evidence Waterman's manipulation of the gay club scene in the 1980s and possibly
call it a sub-genre of music that he then turned into mainstream pop.
By the 1990s though, Watkins was clearly using the same club circuit
to test records that were already designed for daytime BBC Radio 1
airplay and the UK pop charts.

1.5 EUROVISION: THE WORLDS #1 SONG CONTEST

I have always been an enthusiast of the Eurovision Song Contest. It
has played an important part in pop music culture since the 1960s,
at which time it was still considered a lucrative and prestigious ceremony and television show to take part in. The large television audience figures that it has attracted for many years are still substantial
(110–200 million audience viewing figures for recent years) and it will
forever be famous for introducing the world to the Swedish pop sensa-
tion, ABBA, in 1974. It is difficult to know where to place Eurovision
in pop music culture, most of what I am talking about throughout
this book would fit into what Bourdieu (1984) would call 'majority

culture', compared to the world of commercial music where the measure and symbol of good pop music is the national sales charts. Eurovision has almost become a pop subculture or sociology of 'aesthetic discrimination' (Frith, 1996) that could equally be applied to other television shows, such as X-Factor, Pop Idol and The Voice. Certainly Adorno (1970) would suggest that this is part of 'low art', but with a live worldwide audience of hundreds of millions it must mean something special to a lot of music fans who not only tune in but pick up their telephones and vote for their favorite song of the show.

Ian Curnow and I entered a song for the UK Eurovision entry in 1993 when the ex-PWL artist, Sonia, sang all eight songs. Our song 'A Little Time' came third but we had a second opportunity to participate as songwriters in March 1995. Music Week announced on 18 February 1995 that a new style Song For Europe, with Jonathan King drafted in by the BBC as the music executive, was looking for the UK's first Eurovision winner since Bucks Fizz in 1981. The format changed back to eight different acts and songs that the UK public would vote for. On the day of filming the live UK final I was asked by Massive Management to attend the rehearsals at the BBC studios and do what I could towards the live vocal sound of Deuce singing with our pre-recorded backing track. It was a difficult afternoon for me, trying to help the BBC sound engineers (who were reluctant to receive any guidance or requests in their control room) to get the best out of the two main Deuce vocalists, Kelly and Paul. We were all asked to retire to the green room to await a stage call so that all finalists were aware of the walk-down procedure from the green room to the stage for the winning act. I thought it slightly strange at the time that the act, Love City Groove, were chosen as the act to show the rest of us (using screens) what the routine would be from the green room. That seemed all the stranger and coincidental later on when they turned out to be the winners! By the time we got to the live evening performance I was already having concerns about the way the show was going and looking like there may already have been a pre-determined result. Ian and I thought that Deuce gave a tremendous performance that was rapturously received by the audience in the television studio. As soon as the voting began on the phones after the live performances I knew something was wrong. Many of us in the audience were outside on our mobile phones casting votes for whomever we were supporting. The Deuce line was continuously engaged throughout the whole process, none of us there at the contest were able to cast a vote for Deuce and yet Tracy Bennett of London Records stated you could easily get through to line 8 for Love City Groove. By the end of the voting and the live final scores we came third (again) and Love City Groove came first.

The following year, producer and songwriter Nicky Graham had an entry for the same stage of the Song For Europe UK finals with his aforementioned BoyBand, Code Red. Exactly the same thing happened to him from the green room, the act chosen to do the winners rehearsal walk, turned out to be the winner that year as well.

1.6 REUNIONS: THE RETURN OF THE MANUFACTURED POP AND BOYBANDS

A recent television show called 'The Big Reunion', initially broadcast by ITV in 2013, has caused a resurgence in popularity for some of the British pop and BoyBands from the 1990s. Possibly this has come about since the successful comeback of Take That in 2005. Acts featured on the first series included 1990s BoyBands such as Five, 911 and Blue. It is a reality television show and we have seen individuals struggle with returning to the limelight as well as success stories leading to the show becoming a live touring performance around the UK. Comebacks and reunions are a tricky social discourse that can fail as well as succeed with a measure of uncertainty due to the sociological factors between the band members and whether they can put past differences and ego clashes behind them. Songwriter and now music manager, John McLaughlin, offers this recent insight:

> **Phil:** Take That is a fantastic example with their comeback and recent success.
> **John:** They're an amazing example and they've been my inspiration for putting Bay City Rollers together again.
> **Phil:** East 17 tried to get together again after the Take That revival and they allowed the television cameras to follow them but it got to 'fisticuffs' and fell apart again.
> **John:** Exactly – if the animosity is still there and people see that. People want a happy story and a happy ending. If they've got a band trying to get back together again they don't want to see them batter each other. People are not stupid, the public is the public and they'll see through that.
>
> (McLaughlin, 2015, personal interview)

Even though Take That have experienced a successful comeback since their reunion, they began with four members, expanded to five members for an album and tour with Robbie Williams in 2010 but have been reduced to three members before their recent album '111' and the tour in 2014 due to the departure of Jason Orange. Gary Barlow is a prolific, creative music-maker who has been motivated by this reunion and has also enjoyed resurgence in his solo career on the back of Take That's comeback success. McCarthy (2012) states that many of the original fans from the 1990s are not happy about the reunion: 'Many of Take That's fans never got over the group being disbanded and to this day haven't forgiven the band and will not buy any of their new music' (McCarthy, 2012, p.52). Gary Barlow has become a national music treasure in the UK, producing the Queen's Diamond Jubilee Concert outside Buckingham Palace in 2012. No other BoyBand from the 1990s has experienced this level of rebirth success.

There is a version of East 17 currently touring nightclubs and holiday parks around the UK; this line-up does not contain Tony Mortimer or Brian Harvey. It is hard for me to comment on why the East 17 reunion did not work but if it did not contain a Tom Watkins style of figurehead then it was likely to fail if my theory is correct. I believe this type of music act requires an external team-leader or entrepreneur to turn creativity into commerce.

1.7 POP MUSIC CULTURE CONCLUSION

Musical value judgments in the manufactured pop and BoyBand cultural genre of the 1990s need to be understood within the circumstances and contexts in which they were created and constructed. Whilst it is easy to say now that BoyBands existed prior to the 1990s, it is clear that the BoyBand phenomenon reached its zenith during the 1990s by the fact that so many BoyBands were formed and achieved chart success. The managers, record companies, media, journalists and eventually the fans drove the social constructionism that led to the BoyBand phenomenon in the 1990s, all of whom had wanted to see the space left by the demise of the 1980s PWL hit machine filled by groups that were a contrast to the faceless Acid House and Rave music of the late 1980s/early 1990s. What other alternatives did pop music fans have in the 1990s? American grunge rock? Britpop bands such as Blur and Oasis? These were never going to fulfill the requirements of the pre-teen pop fans that wanted to place posters of their pop heroes on their bedroom walls and sing along to a catchy, commercial pop song. The space was there in the field and artists such as NKOTB, Take That and East 17 were manufactured to fill it. Frith (1996) states that high and low culture judgments need to be made through the mutually defining discourses of art and we need to continually refer to the ideological baggage that the individuals in this field carry around with them before we make too many assumptions. The commentary from respondents in this chapter has highlighted Frith's points. For instance, when manager Tom Watkins describes East 17 as 'rough trade' he is comparing the band members to male prostitutes on the gay culture scene of the 1990s. That highlights an important view of how managers were marketing their BoyBands during this period and we will hear more of the Watkins style of cultural imperialism and his style of total project control in the next chapter. Adam Krims (2003) cites an Adornan vision of 'seemingly endless worldwide proliferation of young boy pop/R&B groups' that came to be, certainly in the UK and Europe throughout the 1990s as I have indicated in this chapter and I will offer further examples of this phenomenon throughout this book. Allan Moore (2003) states that scholars have yet to agree on theoretical paradigms for the analysis of Western popular music. Despite the growing number of studies (Tagg, 2012; Frith, 1996; Moore, 2003; Burgess, 2014; Zagorski-Thomas, 2014), few mention the manufactured pop and BoyBand phenomenon of the 1990s. I will present a number of new theories throughout this research that relate to manufactured pop music and the BoyBands of the 1990s. Simon Frith (1996) states that consumers in the 1990s made their own market choices but he underestimates the role of what I consider the real *power* people with the novelty within the (pop music) cultural domain i.e. the entrepreneurs and team-leaders such as Tom Watkins and Pete Waterman. They were the type of people that influenced the consumers' choices by manipulating the music industry system of the day, from the songs creators through to the final product's commercial presentation. I call this a 'Service Model For (Pop Music) Creativity and Commerce'. The system requires a figurehead that has economic, cultural and social value, Tom Watkins was a good example of this in the 1990s, following his 1980s success with Pet Shops Boys and Bros. Watkins constructed the BoyBand *field* through his Massive Management company, acquired the creative *agents* to enter the *field* (artists, songwriters and producers) and through his total control of the creativity, business and marketing, created economic value in the projects. The next chapter will commentate on the creative and commercial schematics of this music genre and its phenomenal success.

NOTES

1. Currently working for online music journal, *Quietus*, and writing the biography of music entrepreneur Tom Watkins.
2. Red coats are the staff at Butlins holiday camps in the UK, where they organize activities and entertainment for adults and children.
3. Tom co-wrote most of the songs on the album but mainly gave his share and credit to Rob Kean.
4. Tom's new musical partner and boyfriend.

2

The Business of Pop Music Production

2.1 INTRODUCTION – INDUSTRY

The music industry's representative body, UK Music, was formed in 2008 by the individual music sector bodies as an umbrella organization to be a music industry equivalent to the UK Film Council. UK Music often refers to the music industry as part of the creative industries and its mission is to be a single voice on behalf of music to the UK government, media and education. Their website headline statement is: 'UK Music is a campaigning and lobbying group, which represents every part of the recorded and live music industry'[1].

A Philosophy of Art?

I will show how the choice of methodology can influence artistic and creative outcomes that then become successful and repeatable commerce. I would argue this: Pop and BoyBands are still essentially a creative phenomenon that whole communities have grown to love, even if they arise from a business process. The cultural dominance of One Direction was highlighted by the fact they were named the world's bestselling artists in 2013[2]. Those who mock the genre simply misunderstand it because they come from different art worlds. This assumption leads them to reason that because popular music is different from art music, it cannot be understood by appealing to prevailing standards of musical value. Gracyk (2001) states:

> A growing number of philosophers regard popular music as a vital and aesthetically rich field that has been marginalized by traditional aesthetics. They argue that popular music presents important counterexamples to entrenched doctrines in the philosophy of art.
> (Gracyk, 2001. Available at: http://www.iep.utm.edu (accessed 29 July 2016))

My aims are to explore this by using today's technology to test my formula framework, working with songwriting teams specifically towards the current pop music market.

2.2 POP AND BOYBAND MANAGEMENT: 'ALWAYS FUCK THE ARTIST BEFORE THEY FUCK YOU'

On the BBC Four television show 'Music Moguls – The Managers' this quote from the deceased music manager, Don Arden of Jet Records and Management, was re-broadcast: 'Always fuck the artist before they fuck you' (Arden, 2016, BBC Four TV. Broadcast 15 January 2016). That had previously been seen on an earlier television show called 'Mr. Rock'n'Roll' broadcast on Channel 4 in 2000. That television show also featured Colonel Tom Parker (Elvis Presley's manager), Peter Grant (Led Zeppelin's manager) and Tom Watkins (East 17's manager). Don Arden had managed the Small Faces in the 1960s and then Black Sabbath and ELO in the 1970s. He was nicknamed 'The Al Capone of Pop' by journalist, Mick Wall, after his death in 2007. Suffice to say that his reputation within the industry was not good. He was known as a tough guy to deal with and rumored not to have paid many of his artists. I had some minor dealings with his label in the late 1970s during an album I engineered at The Marquee Studios and although the project had many problems surrounding it and was never released, everybody at least got paid. The BBC Four Music Moguls program went on to state that there are three main music management categories, with the following examples:

1. **Manipulative Management:** Larry Parnes, managed artists such as Billy Fury and Tommy Steele in the 1960s.
2. **Monetary Management:** Colonel Tom Parker, Elvis Presley's manager.
3. **Artist Protection:** Peter Grant, Led Zeppelin's manager.

I would generally agree with these categories, though I would make an argument for 'Service Management', which is relevant for established artists who already have a history and possibly a long career in the music industry. That type of artist would just need an effective administrative style of management because the media and concert avenues are already available to the established artist. Rod Stewart and Elton John are good examples of this category where they may have started their careers with a manager in categories 1, 2 or 3, but have since parted with them. Some interesting headline quotes came from some of the well-known managers interviewed in the same BBC Four program:

1. Malcolm McLaren (speaking in 2002): 'Don't fear failure. Better to be a flamboyant failure than any kind of benign success.'

2. Jonathan Dickins (Adele's manager): 'I think you have to embrace how things change; all this "stuck in the past" or "that's how it used to be" is not good.'
3. Simon Napier-Bell (Wham's manager): 'We still have to be the same mix of therapists, friend and sometimes parent.'
4. Ed Bicknell (Dire Straits' manager): 'To me what management is about is to take the art, if that's what it is, and turn it into commerce.'

('Music Moguls – The Managers', BBC Four TV.
Broadcast 15 January 2016)

These are just some descriptions of what a music manager's role is, but I do not think any of them capture the role of a manager in the manufactured pop and BoyBand genre. Probably a combination of categories 1 and 3 together with the Larry Parnes manipulative management style would be close to the requirements. Generally, the 1990s managers that manufactured the artists I have commented on in the artist section, such as East 17, Boyzone, Take That and so on, have all started with the management and their brainstorming ideas of the type of pop band they want to formulate. East 17 are an exception to the rule as they started with Tony Mortimer approaching Tom Watkins with his song demos, followed by Tom's advice and feedback to come back with some 'like-minded' friends. The other artists (especially BoyBands) have by a large majority attended auditions and been chosen, styled and groomed by the managers. Louis Walsh (2007) calls it 'Audition Hell; it is hard being judged and it's hard being rejected, but it's part of this business and it's certainly part of the audition process' (Walsh, 2007, p.25).

Nigel Martin-Smith

Here is what East 17 manager Tom Watkins had to say about Nigel Martin-Smith during my recent interview with Tom:

Phil: Did you deliberately drive the rivalry stories between East 17 and Take That in the media?

Tom: God yeah, I think I invented half the stories anyway. There was a true story, I lived next door to Lulu [in Maida Vale, London] and Lulu was singing on one of the Take That records ['Relight My Fire'] and Nigel Martin-Smith was round visiting Lulu one night and I don't think they imagined I was in because I saw Nigel Martin-Smith peering into my Georgian house and swooning. I was on my balcony and shouted 'Oi, you nosy bastard'; I think he nearly shit himself. No disrespect to him, he just wasn't in my league.

Phil: Did you ever get friendly with him at all?

Tom: No, there was no competition as far as I was concerned [between Nigel and Tom]. Nigel Martin-Smith didn't have the ability to design a sleeve, didn't have the ability to hire a designer. All you would hear him say was I told the boys they can't have girlfriends but I turned round and said they should have girlfriends, they should be seen with girls because it's good being seen sniffing a few girls' knickers and all the rest of it and boys underwear for that matter.

(Watkins, 2014, personal interview)

Tom Watkins

We get an idea there of Tom's unforgiving yet totally self-confident attitude. As opposed to describing Tom as a manipulative manager, he was in fact a very controlling manager, overseeing and often creating the many aspects of driving a BoyBand to success. Here are some more examples of Tom's unique management style:

> I would like mentioned in my obituary that I am a control freak and I believe that every time I take full control it happens [success]. The minute I relinquish any control, I think it's very dangerous. You just push me once over the edge and that's it because I was holding and watching every single aspect. It only worked when I was in my professional environment working with Gilberts [accountants], working with Paul Rodwell [lawyer], working with Neil Ferris [promoter] and there was a perfect understanding.
>
> **Phil:** On the marketing and promotion, whether it was external or within the record label, is it right that they would not do anything without your permission?
>
> **Tom:** It was absolutely forbidden.
>
> **Phil:** From the videos to the promotion campaigns, you had control of everything?
>
> **Tom:** That's why I formed my own film company, I formed my own graphics department, I employed my own graphic stylist and everyone else because I simply wouldn't trust people. I have often been quoted as saying it [control] is a 'necessary evil' right, because I knew how important it was once you had those quality goods. You do it for money and all the rest of it but to me it was a secondary thing [money], getting it right was far more important.

(Watkins, 2014, personal interview)

From Tom's commentary, I would highlight the team of business people he surrounded himself with. He would delegate these jobs to people that had the right skills, allowing Tom to concentrate on what he did best, steering the creativity from the songs, to the production through to design, styling and promotion. This underpins the statements that I have highlighted elsewhere, that every link in the chain needs to be strong to turn creativity into commercial success. Clearly this level of control that Tom describes was important to him throughout the 1990s. In the early days of Tom's management of Bros in the 1980s there was a similar amount of control but that was relinquished when Bros demanded control of the songwriting for their second album. The songs were not good for that second album and no matter how hard producer Nicky Graham worked, as Watkins has said; 'You can't make chicken soup out of chicken shit'. I believe this caused his attitude to change from East 17 onwards and he decided to take full control of everything as though he had learned his lessons from Bros and was determined not to make the same mistakes. It would appear that Tom's relationship with Pet Shop Boys was entirely different though, as journalist Matthew Lindsay alludes:

> **Phil:** How much manipulation and styling of Pet Shop Boys did Tom Watkins have compared to Bros and East 17?
>
> **Matthew:** Very little. Tom was the big bad man that went to the record label, thumped his fists on the desk and got the money for Pet Shop Boys [for recording and promotion]. If Pet Shop Boys were informed by Tom it was more as a reaction against what Tom was trying to put across. For instance, for their 'Please' album cover, Tom turned up to the studio and showed them a mock-up of this huge foldout sleeve with maximum content. They balked at it and ended up with a plain white sleeve with a picture of themselves the size of a postage stamp on it. They would say it was a very creative period though because Tom was someone they could constantly react *against* artistically. Tom wanted dancing girls in one of Pet Shop Boys videos and Neil was mortified that he was expected to stand there with dancing girls behind him – so it didn't happen. Pet Shop Boys did actually put showgirls in their video for 'What Have I Done To Deserve This?' (1987). But as with so much of the Pet Shop Boy's best work, their use walks a fine line between passionate commitment and a distancing irony. There's a genuine love of pop music, of showbiz theatrics but it's coming out in a slightly arched way, like a marriage between pop and anti-pop.
>
> (Lindsay, 2014, personal interview)

Again, an indication of Tom's frustration at not being fully in control, which led to a parting of ways with the Pet Shop Boys and I believe fueled his desire by the 1990s to be in full control of the artists he managed. Tom's self-assurance was always something to behold but Lindsay has stated he believes Tom to be 'incredibly insecure' underneath the exuberant and massively confident persona he portrayed. Producer/manager working relationships are strange and difficult to explain at the best of times. The relationship does need to be a complete commitment from both sides, especially from the producer's point of view. It is no good for producers to feel that they want to go one way or work with specific artists when the manager wants the producers to work on something else. It was clear at the early point of our relationship with Tom Watkins that there were going to be conflicts of interest, not least the fact that Tom was managing East 17 (and his other Massive Management artists) and P&E. I knew that at some point in the future Tom would have to choose between one or the other and his choice was always more likely to fall down on the side of the artist; Tom was always willing to challenge Ian Curnow and myself.

Louis Walsh

Here are some examples of what can happen behind the scenes in a producer/manager relationship. When Ian and I met Louis Walsh for the first time at The Strongroom he said that he had been trying to get in touch with us for some time through Tom Watkins and that he admired the work we had been doing with East 17. He had wanted us to work with Boyzone as soon as they were formed but Tom Watkins told Louis that they could only work with P&E if he gave 50% of the management of Boyzone to Massive Management. Little did Ian and I know then that we were about to embark on a two-year journey of working with Louis Walsh and his artists. Whilst Tom Watkins was managing us he also berated Simon Cowell heavily as Simon had dared to phone us directly at our studio to talk business. I believe that may have scared Simon Cowell off from working with us for some time. Simon was another person who approached P&E once our relationship with Tom Watkins finished in October 1995 and we did some initial development work for his BoyBand Five. Ian and I found Louis Walsh's style of management very different to Tom Watkins. To start with he had no desire to manage us as well and he was a lot less controlling. Going back to my earlier descriptions of management styles I would place Louis Walsh somewhere between the 'artist protection' and 'monetary management' categories. Louis always had the best interests of his artists as his priority, as we see here from Nicky Graham and Deni Lew, producers of 1990s BoyBand, Code Red:

> **Deni:** We [Code Red] never got the proper push here in the UK.
> **Nicky:** We were doing the *Smash Hits* [Pop magazine] road tour and any band on the tour that won the award for best new act of the year were basically guaranteed success on the UK pop market.
> **Deni:** There wasn't one new act winner that didn't go on to become successful; N-Sync and The Backstreet Boys are good examples. The only act we know of that wasn't a success was The Carter Twins and they beat Code Red to the new act award, the same year we are talking about [1997], they were Louis Walsh's act.
> **Nicky:** and Boyzone [Louis Walsh's main act] were headlining the tour.

Deni: We won [with Code Red] hands down because how we know is – come the tour – there were three bins at the end of the show for three different acts up for the newcomer award and the fans would put their vote into one of those bins, our bin for Code Red, every night, was full! I'm not exaggerating the other bins were barely half full – our bin was completely full. We won hands down and we were saying to the tour organizer, 'we are going to win this', we can see we've won [by the bins] and yet The Carter Twins won. Nobody liked The Carter Twins, but everybody liked Code Red. When Polydor saw that Code Red had lost they decided not to push the band any further in the UK.

(Graham & Lew, 2015, personal interview)

Ian and I were working with The Carter Twins at that time and we were blissfully unaware of the events taking place. The Irish BoyBand, OTT, were signed to Epic/Sony Records in Dublin by John Sheehan and were also guided by A&R supremo, Rob Stringer, in the Epic Records London office; we had worked with Rob on the 2wo Third3 project a year or so earlier with Tom Watkins. Whilst Ian and I were visiting Dublin during 1996, Louis took us to the Epic/Sony Ireland office to meet with John Sheehan and discuss the project. It was clear to Ian and I that Louis was doing more than just 'helping out' as he was involved in choosing the material and hiring us to produce. He may not have been present in the studio as he was at times with the Boyzone sessions, but his opinion was strongly relied on by the label throughout the project. The first OTT single was a cover (a traditional BoyBand starting point) of The Osmond's record 'Let Me In'. It gave OTT a UK top 20 single by the summer of 1996 and allowed Ian and I to write many songs for the album, producing those and a string of singles throughout 1996 and 1997. Matthew Lindsay and Tom Watkins gave these comments about Louis Walsh during my interviews with them:

I think Boyzone were just too sweet and clean but it's all about your audience. Louis Walsh obviously saw Boyzone only that way and yet he says that David Bowie was his major influence, so you wonder what Louis is doing – has ever done – to permeate that influence in his acts.

(Lindsay, 2014, personal interview)

The fact is I don't think you are to be thanked in this business. When you imagine, if you put me against all my contemporaries like Louis Walsh and all the rest of it, we pissed all over them for what we did. Louis Walsh says in his book if he were ever managed by anybody with a BoyBand, it would be me. Louis has frequently said to me that 'nobody came up to what you were doing'. Louis' attention to detail was never as good as mine; to me the devil is in the detail.

(Watkins, 2014, personal interview)

This highlights the different styles of management between Tom Watkins and Louis Walsh. Lindsay reports that David Bowie was Louis Walsh's biggest influence but in terms of cover songs for BoyBands and pop acts he always seemed to defer back to The Bee Gees. Our relationship with Louis Walsh and Boyzone went into rapid decline in 1997 when the A&R decisions for Boyzone's third album shifted from Polydor Records, Dublin to Polydor, London. Louis had already asked us to consider producing the first planned single for the third album, a cover of Tracy Chapman's 'Baby Can I Hold You Tonight'. That job was handed over to Steve Lipson, who was also managed by our then management, Zomba. We were a little disappointed that Zomba did not choose to fight our corner on that decision but that is often the way for the larger producer management companies; it is hard for them to side with one of their producers against another, they would always consider it best to give the client what they want.

Steve Gilmore: Management vs A&R – 1998/1999 Mero Project

Ian and I had already collaborated with songwriter John McLaughlin on a number of pop projects, such as BoyBand, Five, for Simon Cowell. John kindly introduced P&E to 911's manager, Steve Gilmore. We then had a short period around 1997–1998 working on products with 911. Steve's next act to launch would be the boy duo, Mero, who had been signed to Simon Cowell's label at BMG Music. Steve Gilmore and Simon Cowell's concept for the band's direction was 'Motown meets Wham'. The Mero project is an example of how fractious relationships can be between A&R and management.

The Mero album was completed for an estimated £250,000 (recording budget) and the first single 'It Must Be Love' was released in March 2000 with a rumored marketing spend of another £250,000. That included two video shoots, the second of which was shot in Miami, USA. The single poses some interesting scenarios that I have often demonstrated to students in universities around the UK. The first version of 'It Must Be Love' was a very professionally produced song demo, co-written by producer David James and lyricist John McLaughlin. Even in the early 1990s, when Ian Curnow and I had signed a publishing deal with BMG Music, we quickly learned that industry professionals, such as record label A&R and managers, expected to hear a song demo produced to a very high standard. No longer were a guitar and vocal demo or a piano and vocal demo acceptable, which came to an end in the 1970s. Looped drums, guide bass, guide keyboard and a rough vocal demo were also unacceptable (this was generally fine throughout the 1980s). Now, only a fully produced demonstration of what the final record would sound like was accepted and expected.

To get a song written from scratch, it would generally be plotted from a defined direction and then produced fully, with drums (programed), bass (programed), keyboards, guitars, lead vocals, backing vocals and brass plus strings (both programed). This would therefore become at least a three- to four-day studio job. That was the case in the 1990s and it would still take that long now, depending on the number of people in the songwriting and production team. Songwriters are expected to do all of this 'on spec', i.e. with no advance payment, no musician costs and no studio costs paid. Only successful acceptance of a song and a commitment to move into full production would allow a production team the chance to charge a fee, usually 50% on commencement and 50% on completion[3]. The song demo of 'It Must Be Love' produced by David James was immediately accepted by Simon Cowell to go into full production, probably for a budget of around £10,000. For the full production version, David replaced his programed brass with a live brass section, added more backing vocals and made the programed drums sound more like a real drummer; he also re-programed the keyboards and bass.

Simon Cowell was not happy with David's initial mix and instead of asking for changes or improvements, he decided to hire ex-PWL remixer Pete Hammond to do a more 'programed' straight ahead pop mix. The drums were re-programed to sound less like a real drummer and more like dance pop, a pop string line was added to the introduction and a 'round patterned' bass was programed in the style of the strings. The brass was lowered in the balance (Simon Cowell does not like real brass much, as Ian and I found out when he criticized a live brass arrangement that we had added to an album track on the Mero project). The Pete Hammond remix sounded more like 'Right Back Where We Started From' by Maxine Nightingale than Motown or Wham. Simon Cowell liked this mix but Steve Gilmore and Mero did not; they had a vision of the two boys on a live concert stage with a full live band, including a brass section. Meanwhile, Ian and I were preparing four other tracks for the Mero album at our P&E Studios at The Strongroom. These sessions were going very well and Steve Gilmore was delighted with our progress, so he made the suggestion to Simon Cowell that P&E could do another remix of 'It Must Be Love' that would be a compromise between the David James version (liked by Steve Gilmore and Mero) and the Pete Hammond version (liked by Simon Cowell). So, for version three of 'It Must Be Love' we removed the pop strings, re-programed the drums and bass again, added some Motown style keyboards and raised the brass in the mix. Steve Gilmore and Mero were delighted with this version, Simon Cowell said it was 'nearly there' but was missing something. We invited Simon Cowell into the P&E studio to see if Ian and I could find out what was missing on this track for him. We recalled the mix on our system and Ian had his keyboard and a whole set of MIDI sounds and samples ready to experiment with. When Simon Cowell arrived he asked if there was anything more indicative of the Tamla Motown sound that we could add? The turning point was when Ian started experimenting with percussion samples on the keyboard, playing various sounds and patterns. It was when Ian hit a splashy tambourine that Simon Cowell got excited and said 'that's what's been missing from this track' and that was it, he was with us less than an hour and went back to his office feeling happy that he had found the 'missing link' from this track back to Motown. The single was finally released in March 2000 and only just reached the UK top 40 charts. This was disappointing for everyone involved considering the recording and marketing budget.

The differences of opinion between Simon Cowell and Steve Gilmore and the direction of the single and the album completion after the events I have described became problematic and it was increasingly clear to me that Simon Cowell was not happy about being challenged by Steve Gilmore on the direction of the act. The feeling Ian and I had as we delivered our album tracks after the single was that Simon Cowell had been increasingly dissatisfied with the whole project. The single's bad performance gave Simon Cowell the opportunity to shelve the whole project immediately. Whether Mero could have gone on to be successful, we will never know. The project needed that kind of recording and marketing budget to succeed and the chances of another label taking over after the first single were very unlikely.

BMG Records terminated Mero's recording contract. Many good songs and productions remain unreleased as other artists and A&R people in the 1990s were not keen to record songs from a project that had already been seen to fail. This is a unique 'insiders' view of an A&R versus management battle and is by no means a representation of other such relationships in the music industry where these two vital roles work together successfully. I asked songwriter John McLaughlin for his views on the outcome of the Mero project:

> **Phil:** Do you think tensions between Mero's manager, Steve Gilmore and Mero's A&R Simon Cowell made any difference to the situation?
>
> **John:** Yes, I think so; Steve's been a friend of mine for years and years. He's very, very strong-minded, very thick-skinned and very single-minded. So is Simon Cowell, so therefore you are obviously going to have fireworks here and there and looking back I'm sure Steve would admit himself that there's only ever going to be one winner in that game. And that was Simon [...]. I got to know Simon over that period of four to five years when I did 'Busted' and Westlife's 'Queen Of My Heart' and a bunch of tracks for Five and I think Simon's one of the most genuine people I've ever met in the industry and one of the nicest people I've met in the industry. I think he genuinely cares about people but if you overstep the mark he maybe won't support you. I think he did his best by those boys and he gave them decent sums of money to have a go. I think for the label, if a project like that doesn't react at radio, they look to move onto the next thing, like a poachers dog.
>
> (McLaughlin, 2015, personal interview)

It is good to see a contrasting view to mine from someone else involved in the same project but we must remember it was Simon Cowell's record label. It is possible, as John suggests, that the executives at BMG Records above Simon Cowell may have inflicted the fatal blow on the Mero project. I have highlighted this particular project to illustrate the kind of character conflicts that exist behind the scenes of manufactured pop acts.

Pop and BoyBand Management: Final Notes

To bring the subject of manufactured pop and BoyBands management up to date for 2019, songwriter and producer Ian Curnow has recently been involved in the development of a new BoyBand and has these comments on the current scene with a reflection back to the 1990s for comparison:

> Most of the people who want to manage BoyBands now are just like tour managers, they've smelt money. It is shark-infested water, full of low life. Everyone tries to pinch everybody else [band members].
> **Phil:** Do we not have managers around now like we had in the 1990s such as Tom Watkins and Louis Walsh?
> **Ian:** No, because Simon Cowell has squashed all of that. Unless you're in with him or unless you are like The Wanted working with a massive management team like Global Management (they also manage Justin Beiber) with a massive machine behind it. But their records [The Wanted] were a bit shit and the band were a bit shit and they're not in profit – they're just making a lot of noise and now they've split up.
>
> (Curnow, 2014, personal interview)

Ian describes the *longevity* situation around The Wanted to be similar to my commentary earlier about Deuce in 1995, where the act were dropped by London Records for not being in profit, even after three hit singles and a top 20 UK album. The current BoyBand management scenario that Curnow describes is similar to the way that the television program 'Boyz Unlimited' (broadcast on Channel 4, 2000) portrayed BoyBand management.

2.3 PRODUCER AND SONGWRITING BUSINESS DEALS

To conclude this chapter I will reflect on the typical contracts and deals for the producers and songwriters working for this genre in the 1990s. Whilst creative producers are always focused on the aesthetics of the songs and productions, we are always drawn back into the fact that this is also a commercial activity. The longevity and survival of production teams in the pop music genre relies heavily on good business negotiations and contracts. The contractual business for P&E Music throughout the 1990s was taken care of by lawyers Simon Long and Penny Ganz at The Simkins Partnership. The business negotiations on behalf of P&E Music were conducted by our managers, Massive Management (1992–1995) and Zomba Management (1995–2000). I will also outline the 2019 working models. Within the producer deals I will also compare the pop music genre to the rock, jazz and classical music genres, as the cross-references are relevant.

Phil Harding: Producer deal Perspectives 2019: Conceptual

Since music producers, such as Sir George Martin in the 1960s, fought to receive a share of record sales royalties, there have been two main types of deals for producers. My descriptions of the producer deals that exist are from my own experience as a music producer in the pop/dance music genres since the 1980s. Whilst I consider these to be 'industry standard' they are not the only types of deals that exist and one should always seek legal advice when entering a production contract. My examples are the main points in what is called a 'deal memo' or 'heads of agreement'; they are considered legally binding and should be signed at the commencement of a production. Typically they consist of two pages forming the basis of what could eventually become a fifty-page producer contract (largely based on the artist's contract with the record company because the producer's royalty is taken from the artist's royalty by the record companies in the UK and Europe), generally completed and signed around six months later, often long after the production is finished and delivered.

Example Deal #1: The Costs Deal – £10,000 Production Budget per Track

Traditionally traced back to the first type of producer agreement that Sir George Martin instigated and still used today in the rock, jazz and classical music genres. Typically, the main contractual points are:

1. The producer receives an advance fee of £2,000, recoupable against royalties. (Meaning that sales to the value of this amount for the royalty have to be earned before more money is paid to the producer.)
2. Royalties paid to the value of 3% of the recommended retail price (RRP). (There can be a pro-rata at a higher rate if calculated to the dealer recommended price – (DRP). Around 4.3% for this example.)
3. All recording studio, session musicians and equipment hire costs to be paid by the record company.
4. Any costs relating to orchestral overdubs to be paid by the record company.

5. A-side/B-side clause for singles. (Meaning the 3% royalty is not reduced if other producers are used for a vinyl B-side/CD bonus tracks/any other digital package including bonus tracks.)
6. The producer royalty will not be reduced by any more than 1% if another producer is hired for remix and additional production work for the final 7" A-side/radio/video version.

Example Deal #2: The All-In Deal – £10,000 Production Budget per Track

This deal originated in the 1980s initially to accommodate the growing breed of superstar DJs who were being asked by record companies to deliver club remixes for their pop artists. Ian Curnow and I worked with this type of deal on all of our pop and BoyBand productions in the 1990s, by which time this type of deal had progressed from suiting the remixers of the 1980s to becoming a new music industry standard for producers in the UK and Europe for the pop and dance music genres. The main object of this type of deal is that administrative and negotiating work is taken away from the record company (they like that as it causes fewer problems). All of this work is handed over to the producers (often with the help of the producer's management team). This means that a producer/production team must be able to negotiate studio bookings and ancillary studio costs, musicians and their negotiated fees. This is a good example of the number of 'hats' that a pop producer has to wear – creator, technician, musician, diplomat, administrator, negotiator – this adds up to quite an array of varied skills. The key to negotiating the all-in deal is to make sure that the full budget is not 100% recoupable against the producer's royalty. Here are the main contractual points for an 'all-in' deal:

1. The producer receives the full £10,000 budget: 50% on commencement, 50% on delivery.
2. Only £2,000 is recoupable against the producer's royalty. (Note that this is the same amount as example 1. Therefore the remaining £8,000 is purely for costs and would be stated so in the contract.)
3. Royalties paid to the value of 3% of the RRP. (Same 4.3% pro-rata for DRP as example 1.)
4. If a full live orchestra arrangement and recording is required or requested (this can cost up to another £5,000) then this will be an extra cost to the record company above the £10,000 budget.
5. A-side/B-side clause for singles. (Same as example 1.)
6. Producer royalty will not be reduced by any more than 1%. (Same as example 1.)
7. Further fees required for mix recalls beyond mix two and for extended and club versions.

Example Deal #3: The Fee-Only Deal – No Royalty/No Contract

This simple deal has developed due to the collapse of physical record sales since the introduction of the MP3 and Napster in the late 1990s/early 2000s. In a market that has shrunk by over 50% in industry revenue it is clearly 'survival of the fittest' in the music production world. Recording budgets, whether record company funded or artist/management/publisher funded, have reduced to around 25% of what they were in the pop genre peak of the mid-1990s. So, if one used to get £10,000 to produce a single pop track in the 1990s, one may now only be offered £2,500 or less for an equivalent production. If you are 'top of the tree' in the music production pop world now, you could still get £10,000 or more, plus a good royalty percentage, especially if you are in demand. But many producers now will prefer to negotiate as high a fee as possible knowing that perhaps only 5–10% of records released worldwide in 2019 are successful enough to warrant the high legal fees to negotiate producer royalties. This keeps the deal negotiating simple for producers and their managers. The tendency now is to charge 50% on commencement and 50% on delivery, no sales royalty.

Remix and Additional Production Deals for Pop Producers of the 1980s and 1990s

Ian Curnow and I were amongst the early wave of pop and dance remix specialists in the mid-1980s. Increasingly during our time at PWL, we would be offered remixes of pop records and asked to create a more commercial and radio-friendly mix in the style of the PWL sound. This would generally require a completely new backing track, often keeping only the vocals from the original production. We asked the PWL business affairs director, David Howells, if he could negotiate a 1% royalty for us, plus a new style of credit that would read 'Remix and Additional Production'. As record companies began to agree to this, it quickly became a standard industry deal as outlined:

1. 1% royalty on retail (this would cause a reduction of the original producer's royalty from 3% to 2% on retail).
2. A-side/B-side clause – see example #1.
3. 50% fee on commencement and 50% on delivery.
4. Further mix recall fees beyond mix two.
5. Further mix fees for extended and club versions.

The USA 'Hoff' Deal: 1990s P&E Music/David Hasselhoff Production Contract

In the mid-1990s Ian Curnow and I, wrote and produced two tracks for David Hasselhoff. Even though David was signed to Arista Records in Germany, where he had had a lot of success as a music artist in the late 1980s, his management insisted that the contractual negotiations would be with David Hasselhoff Productions Inc. USA and not with Arista Records in Germany. Our lawyers at the Simpkins Partnership in London did their best on our behalf to persuade David Hasselhoff's management that we would prefer to be contracted directly to Arista Records in Germany but they refused. When the contract arrived from America we were shocked to see the differences between this and a standard UK/European label production deal. The major difference was that we would see no royalties for our two productions until 100% of the recording costs for the whole album were recouped. We had no control or even knowledge of the rest of the album's recording costs or production arrangements. Our concern for this type of deal was that recording costs could be run up to high levels by other producers on the project and we would never get a full understanding of the point at which full recoupment would be reached. We did eventually receive some royalty statements from America and suffice to say they did not include any further payments beyond our original production advance fees. We had never come across this type of deal before but were assured by our lawyers that this was standard practice for producers in America at that time in the 1990s and is still applicable today.

2019 Update

The three previous producer deal descriptions are very much from a working producer's perspective, however, in any project that a producer may be involved with, further advice will be required from a music industry lawyer, ideally one that has experience with music producer deals. In 2019, even the simple Deal #3 example now seems to be fraught with problems around self-funding artists and record labels wishing to redefine what we call the back end payment i.e. the delivery or completion of a production project. Most producers and their managers might refer back to my examples and consider them to be industry standards that have been with us for many decades. If we elaborate further on the Deal #2 example (all-in deal), it has generally been accepted by the producers and their clients since the 1990s that the first mix delivered is very much a 'listening copy' for all those involved, i.e. artists/managers and record company A&R. Studio and music production technology since the late 1990s has allowed production teams to completely recall mixes at the touch of a button. This has allowed adjustments of mixes to be a fast process after the first mix and comments coming back to the producer are usually for mix adjustments such as louder lead vocals, and so on. At this point the second mix is created and the requested adjustments, which could even include adding some new production parts, will be delivered again to everybody involved. Delivery of this second mix has traditionally been considered the time to present the second invoice for the 50% completion/delivery payment. Any further adjustments required after this would traditionally incur further production time and studio costs, generally charged on an hourly or daily basis. This point would have been added from the mid-1990s onwards to the producer agreement described in Deal #2.

Since around 2005 this has now turned into self-funding artists and record labels expecting anywhere between two and ten mixing and production adjustments until they consider the completion fee is due for payment. There has even been some discussion in recent years of USA record labels stating in producer contracts that the completion fee will not be paid until the commercial release of the product. That could take months or even years after the productions are delivered, creatively and technically. Producer managers and lawyers are contesting this as it could lead to late or delayed completion payments or, at worst, no second payment whatsoever.

In the five to ten years leading up to 2019, other types of deals have increasingly crept into producer contracts, especially with new and developing artists. It has become commonplace for producers and their managers to participate in the artists' songwriting and publishing rights. This will apply even if the producer does not contribute a musical note or a lyric to the songs. This should be agreed and signed off before the recording dates as it will be difficult to negotiate during or after the recording sessions. This raises an interesting question of where one draws the line between production and a producer collaborating on the songwriting with or for the artist. Since recorded music started, recording engineers and music producers have been making song production suggestions and decisions that border on songwriting changes and input. One could argue that changing one lyric line for an improved production flow of the record is immediately classified as co-songwriting.

Producers in the rock, jazz and classical genres will nearly always make song arrangement suggestions, such as bringing an 8 bar bridge down to 4 bars or a 16 bar chorus down to 8 bars; one could construe that as songwriting input. When a producer is dabbling with arrangements, lyrics and even chords and melody, the songwriting claim will be even stronger. Seldom will it ever be given or won though unless this is negotiated in some way in advance. Those advanced negotiations should be between the producer and artist or between the producer manager and the artist manager. The song publishers of the artists will also need to be notified of any songwriting splits with the producer.

Producer/Manager Deals

Many successful music producers and engineers will have management companies to represent them. There are, in my experience, two different ways of financially structuring a producer/management company deal:

1. *The industry standard.* The management company negotiates all fees and contracts and receives all funds from production fees and royalty earnings, including receiving the producer royalty statements from the record labels. The management company accounts to the producer after retaining their management commission, generally 20%.
2. *The producer-controlled deal.* The producer receives all fees, royalty earnings and statements but the management company negotiates on behalf of the producer for all fees, contracts and royalty earnings with the record company. The producer decides whether their lawyer deals with the contractual negotiations or the business affairs representative at the management company. Finally, the producer then accounts to the management company for their 20% commission of fees and royalty earnings.

Producer/Manager Relationships

Since the 1980s/1990s there have been a growing number of music industry managers and management companies in the UK that will specialize in managing music producers and engineers. Some may also represent professional songwriters, many of whom will also produce their compositions. Some of these managers will come from the experience of managing artists successfully and may still manage some artists as well as producers. Often the larger commercial recording studios in London will add a producer and engineer management department or company to their business. Business orientated recording studio owners and managers realize that many clients booking the studio would ask the studio bookings manager for advice about hiring music producers and more often freelance music audio engineers. This puts them in an advantageous position to offer the music producers and engineers that they represent.

In 1992, Tom Watkins quickly realized, after Ian and I successfully remixed 'House of Love' by East 17 and it became a UK and European hit, that he would benefit greatly in his vision of building a bigger team around him at Massive Management if he managed a producer team that could provide the productions for his growing roster of artists. There was clearly the potential for a 'conflict of interest', which eventually grew over the three years that P&E were managed by Tom Watkins. By the summer of 1995 that conflict would be severely tested during the making of the third East 17 Album 'Up All Night'. A meeting was called to discuss the songwriting splits and the production deal for this third East 17 album where Ian, Rob Kean and myself had been asked to produce the whole album. The first thing Ian and I asked for was a producer royalty of 4% on retail rather than the usual producer royalty of 3% that we had had for our contributions on the first two East 17 albums. This was refused by Massive Management, as was our suggestion to share the songwriting equally between the four band members and the three producers. The inevitable outcome of this was that Ian and I decided to take on the production work on the third East 17 album on whatever terms could be best negotiated for us by our lawyers, deliver the album and terminate our relationship with Tom Watkins and Massive Management. Our view was that we would prefer to find a new management company that would work *for* P&E, as opposed to it seeming that we were working for them.

At some point before this it was agreed between P&E and Massive that if Tom Watkins was going to commission our publishing earnings then he would not be involved in the songwriting splits, whether or not he was co-writing with us and Rob Kean. Tom was clearly not happy with this but we insisted on no commissioning of our publishing royalties, although we did agree to the extra commissioning of our songwriting performance earnings. This is a good example of the complex levels of business arrangements between producers and their managers, especially when the managers are involved in managing the artists and even the songwriting process as well. I should remind the reader that no management contract existed between Tom Watkins and P&E Music and any changes in the original 'gentleman's agreement' would be negotiated verbally in meetings.

Tom Watkins would then trust that as royalty statements arrived at P&E we would copy those onto him and request an invoice for the appropriate commissioning fees. This became particularly complex with performance income statements where P&E would have to separate the Tom Watkins commissioning earnings from others. Eventually P&E had to hire a third party royalty auditor to do this job twice a year and then deliver professional independent statements to Massive Management. P&E still pays that commission to this day, though it has become simpler as royalty earnings have significantly reduced twenty years later.

Pop and BoyBand Songwriter Deals: P&E Music/BMG Music Publishing Deal 1992

There was a cultural change in publisher and songwriter contract terms by the 1990s, having moved from traditional 50/50 splits in the 1950s to 1970s, then on to 60/40 splits in the 1980s, in favor of the songwriters. We soon saw 70/30 splits becoming the standard in the 1990s with an option to move onto 80/20 by the end of the decade and ongoing.

The 1992 BMG Music publishing deal offered to P&E Music contained a commitment that we would write a certain number of commercially released songs to fulfill a three-period exclusive songwriting contract. P&E received a good financial advance for this but almost a year into the deal Ian and I had begun to realize that we were far from fulfilling our commitment. Most people would naturally try to fulfill these three period commitments by considering each period equal to one year; that is how Ian and I optimistically viewed the contract. At the time of signing, our lawyer who negotiated the contract for us, Simon Long, and the BMG managing director at that time, Paul Curran, had made it clear that a period could equal three years if the song release commitment was not met. So potentially on the negative side for us, this could turn into a nine-year deal if P&E did not get the commercial song releases required. Ian and I therefore set aside time to achieve these targets. Hit productions were not enough without an involvement in the songwriting in the 1990s and that is still the case now for pop producers.

In 1996 an administration deal with BMG Music was negotiated to allow our assistant, Julian Gallagher, to join the P&E publishing deal with BMG on co-compositions that we were increasingly offering him and to be able to offer the same to other co-writers we were working with. The administration deal also provided some funding to offer (after approval by BMG) on an ad-hoc basis to any new artists that were signed to P&E Music Productions. This was all part of a general re-negotiation of our P&E Music publishing deal with BMG Music, having fulfilled our commercially released song obligations by 1995/1996. We had also fully recouped our original 1992 advance against publishing royalties and were now in profit. By 1996, BMG Music wanted to retain the same contractual splits and agreements and simply pay P&E a new advance to renew for another three periods. Our view was that P&E did not need the advance money; Ian and I preferred to adjust our publishing share from 70/30 to 80/20 in our favor (and backdate this to 1992). It was an easy deal for BMG Music to agree and P&E received everything that was asked for, largely thanks to the negotiation skills of lawyer Simon Long of The Simkins Partnership. This is an example of how producer/songwriter publishing contracts can develop once a period of success is sustained. The examples of the songwriting success P&E achieved with East 17 and Deuce that I have reflected on earlier in this chapter created the platform for these negotiations.

2.4 BUSINESS OF POP MUSIC PRODUCTION – CONCLUSION

This chapter has given a broad overview of the music industry aspects affecting my core subject. With the aid of my interview participants I have reflected on the inception of the creativity, starting with the songs, through to the production aspects, the artists themselves, the managers and finally, the types of business deals for all parties involved. The oral contributions from my interview respondents have helped to underpin my own suggestion that there existed a repeatable formula for this genre of music in the 1990s and that we can ethnographically update this for today. The cultural discussions throughout Chapters 1 and 2 highlight the manipulative world that surrounded the manufactured pop and BoyBands of the 1990s and how that was viewed from both parties, with some honest and enlightening commentary by East 17 manager Tom Watkins.

Management

Keith Negus (1992) states that managers such as Tom Watkins, Larry Parnes and Maurice Starr (NKOTB) are portrayed as 'starmakers and svengalis' and he considers them a minority amongst other artist managers who will be more representative and guiding, as I have indicated in this chapter. I would suggest that this manipulative style only really works for manufactured pop acts and BoyBands and my respondent, journalist Matthew Lindsay, has underpinned that view and has even suggested that other managers, such as Malcolm McLaren, should be remembered for the same level of manipulation and grooming with The Sex Pistols in the late 1970s. Negus (1992) also states that the commercial marketplace is not always 'out there, it has to be made' and the commentary from Tom Watkins and his own admission of his controlling style of management gives a strong indication of how far the instigator of a manufactured pop act of the 1990s had to go. My commentary on the management verses A&R scenario around the pop duo Mero shows how an artist's career can go into free-fall when there is not a clear 'team-leader' within the ranks. The alleged 'manipulation' displayed by Boyzone and The Carter Twins manager, Louis Walsh, to help The Carter Twins win the Smash Hits Tour 'Newcomers Award', further underlines the extent that a successful manager will go to 'service' their artist and record company.

Industry and Business

The aim of this chapter is to outline every aspect of the types of producer deals from the 1990s with an update to today's working practices[4]. Typical songwriter deals and working practices of the 1990s are also featured in this sub-chapter; this era was the start of the rise of the producer/songwriter as a working standard. This highlights the kind of diversity that grew out of a necessity for creative music industry practitioners to diversify their skill bases. We saw producers and engineers that may not have collaborated on songwriting before, enter into this field as well as session singers expanding into 'topline' and lyrical songwriting collaborations. Other types of session musicians also began to broaden their horizons from the late 1980s onwards, such as drummers diversifying into drum and rhythm programing as a service to producers and songwriters. The other rise we saw in the early 1990s was the proliferation of managers and management companies specializing in representing producers, engineers and songwriters.

NOTES

1. Under that UK Music umbrella are bodies, such as BASCA (British Academy of Songwriters, Composers and Authors), MPG (Music Producers Guild) on behalf of music producers and engineers, plus the FAC (Featured Artists Coalition) on behalf of music artists. The business side of the music industry also has its individual trade organizations, such as The MMF (Music Managers Forum) representing the interests of music artist and music producer managers. http://www.ukmusic.org/about-us (accessed 8 July 2016).
2. Statistics published by the BPI state the X Factor-originating BoyBand cut a swathe through most territories. Their third album Midnight Memories – released on 25 November, selling 4m copies around the world in just six weeks – was not only the biggest seller in the UK, but also topped the global albums chart. https://www.theguardian.com/music/2014/aug/01/one-direction-named-the-worlds-no-1-bestselling-artists (accessed 21 July 2016).
3. See the producer business deal examples in section 2.3.
4. I submitted the 'Phil Harding Producer Deal Perspectives' section to lecturer and writer, Paul Rutter, author of 'The Music Industry Handbook'. The 2nd edition of this book was published by Routledge in 2017 and featured my contributions.

3

Pop Music Production Creativity

3.1 INTRODUCTION

This chapter focuses on creativity, specifically within the various stages of music production. Many academics have written widely on the general scale of creativity in the arts. Some call it 'The Myth of Artistic Creativity' (Weisberg, 1986) as though artistic genius can never really exist and is always explainable in some kind of scientific or cynical way '...the composer as borrower' (Weisberg, 1986, p.130). I will talk about the 'artist as borrower' throughout this book as it is a commonly used method of pop songwriting and composing but it is within the aspect of '...the artist who imitates aspects of another's work but never actually copies it' (Ibid, p.110).

Mihaly Csikszentmihalyi's 1992 book, *Flow: The Classic Work on How to Achieve Happiness*, is amongst the best examples of explaining creative *flow* in the workplace and I refer to this work many times throughout this book. All producers and engineers in pop music production set out to create a novel work and where possible, put an individual stamp on what they produce. Most music producers and music engineers in 2019 consider themselves to be creative technicians. In order to survive in the music and media industries, virtually all of them have had to diversify and learn new skill sets to continue their careers. Therefore, creativity is at the heart of music production in 2019 and cannot be ignored. Many excellent creative systems and models (ANT[1] & SCOT[2]) have been presented through academic research by academics, such as Simon Zagorski-Thomas (2014). The discussion in the academic world around creativity in music production has become focused on a *non-linear* approach, aligned to Csikszentmihalyi's (1988, 1999) System Model of Creativity. This model has been revised by Susan Kerrigan (2013) and analyzed in detail by Philip McIntyre (2012), McIntyre, Fulton & Paton (2016), Paul Thompson (2016) and Justin Morey (2016).

'Coupled with the research of European empirical sociologist Pierre Bourdieu (1977, 1990, 1993, 1996), we believe that this way of seeing creativity supplies the most comprehensive attempt so far to explain creativity as a system in action'.

(McIntyre, 2018, p.23)

Whilst I agree with many of their *non-linear* views and theories on creativity, I still have the view that many aspects of music production creativity are *linear*. Much of my creativity commentary throughout this book also reflects the work of sociologist Bourdieu and I would recommend readers refer to the excellent work, *The Creative System in Action* (McIntyre, Fulton & Paton, 2016). Sawyer's 'Zig-Zag, 8-step' (2013) path to creativity aligns with some of my own creative practices such as the '12-step' approach to pop mixing, described in Chapter 6. This chapter sets out to highlight these *linear* aspects in pop music production and songwriting as an alternative to those that consider creativity to be entirely *non-linear* in music production. This will be explained and highlighted in detail throughout the following chapters. A vital factor to successful creativity in pop music production is routine and scheduling, as Sawyer (2013) points out from Csikszentmihalyi's research:

'Mihaly Csikszentmihalyi has found that exceptional creators, from all walks of life, schedule daily idea time. No matter how busy they are, they guard their idea time religiously. The reason's simple; you'll have more ideas if you devote more time to having ideas'.

(Sawyer, 2013, p.149)

I was lucky enough to attend a one-to-one interview with Benny Andersson of Abba fame at the Art of Record Production Conference 2017, in Stockholm. He surprised many of us in the audience by saying that at the age of 71 he still practices on the piano for hours every day. When asked why he still did this, his reply was:

'You have to get rid of all the rubbish to find out 4 bars where you feel it – it's not like you think it, it's a physical sense of something good and once that happens I hang on to those 4 bars or 8 bars and then I can start to develop that into a full song. I know that if I don't sit there for hours every day it's not going to happen for me, but feeling it also means you remember it'.

(Andersson, 2017)

Finding and dedicating time for music production creativity is important, especially 'idea time' because ideas may not necessarily come when needed, under pressure, in front of a music production screen. If one can develop a daily creativity schedule that suits, allowing time for new ideas, critical reflection and so forth, then it will be possible to build a pool of creative ideas and resources to call on during the daily workflow. Innovative companies, such as Apple, Microsoft and Google, will encourage employees to spend 10–20% of their work schedule 'thinking up new ideas'. Whether it is thirty minutes each morning of free writing into a journal or a period of time later in the day listening to music references (see Chapter 4) and dictating ideas into a phone. Try to find a solution that works. These notions and routines seem especially important for fiction writers so why shouldn't pop songwriters and producers practice them as well? Some music artists, such as Alex James of Blur, view creativity differently from pop music producers:

> 'I don't think you can manage creativity. It's irrepressible, it's indomitable, it's like thistles, and it just springs up everywhere. You can't stop it and you can't manage it you just have to make sure you're pointing it in the right direction'
> (James, 2018, BBC Four program 'Hits, Hype and Hustle')

Pop music producers, by contrast, will spend the majority of their time trying to manage creativity but will also do their upmost to point it in the right direction. With the aid of my interview participants I will reflect on the inception of pop music production creativity, starting with the songs, through to the production aspects, mixing and the artists themselves. The oral contributions from my interview respondents have helped to underpin my suggestion that there existed a repeatable formula for this genre of music in the 1990s and that we can ethnographically update this for today. The cultural discussions throughout this chapter and the following chapters will highlight the manipulative world that surrounded the manufactured pop and BoyBands of the 1990s and how that was viewed from both parties, with some honest and enlightening commentary by former East 17 manager, Tom Watkins.

3.2 PRE-PRODUCTION CREATIVITY

My own pop production techniques have been described as 'systematic' (Lefford, 2018) and a large percentage of the creativity in that system exists in the pre-production planning stage. If I can't *visualize* how the finished production or remix will sound at the early pre-production stage then I have learned that I will struggle throughout the rest of the production processes. My view now is, if you can't formulate a *sonic picture* in your mind of the final record, then do not complete the pre-production process and turn the project down. Therefore, pre-production starts before accepting a project; it is part of that vital decision-making process.

However, once the production team, guided by the team-leader, has made the decision to proceed with a project and can visualize the sonic picture, the pop music creativity can begin and hopefully it will be a fast, innovative process with ideas flowing freely and sometimes manically. It is important at this stage to be clear of and to have re-sourced any reference tracks that the team leader (generally) will have suggested for the relevant music parts (bass, keyboards, drums, and so forth) and the relevant arrangement ideas (chorus lengths, for in-stance). As a team leader, these will be useful to present to your pro-duction team musicians and programers. Connecting these various references will lead to synthesizing something new, novel and crea-tive. Generally, pop music is programed on computers using software such as Logic Pro and Pro Tools, rigidly referenced in a mechanical way but in the best examples, fluid, flowing and containing original creativity.

3.3 CREATIVITY IN POP MUSIC PRODUCTION

When a pop music programer/producer sets out on their journey to write a new pop song or program for an existing song, production or remix, they will generally start by listening to the references supplied by the client or the team leader of the production team. They could be the team leader themselves or working alone towards a new con-cept but 90% of the time they will be looking at reference tracks, also known as bullseyes or plots in the pop world, to base the new produc-tion, remix or song around.

Music production programing, in my view, will need to involve a minimum of three people:

1. Team leader.
2. Music programer and arranger.
3. Rhythm programer and arranger.

If a new song is required or requested a top-liner/lyricist will also be needed as the other three team members will not be specialists in writing lyrics and top lines for vocalists. More often than not, the best choice for a pop production team top-liner will be a vocalist who has the experience or willingness to collaborate in the pop music genre and is willing to record a guide demonstration vocal with backing vocals to promote the use of the song to clients.

3.4 CREATIVITY WITH SESSION MUSICIAN-DRIVEN POP MUSIC PRODUCTION

Due to vastly improved speed and quality of technology in the twenty-first century, professional musicians need to have a working knowledge of music technology in order to participate in recording sessions. The cost of music technology equipment has dropped dramatically since the turn of the century, allowing more people to participate in music production and recording in their own homes and spaces. Since the introduction of file-sharing platforms, such as Dropbox and WeTransfer, pop and dance music producers expect that the musicians they hire will be able to accept stem files of the productions they are asked to contribute to. The session musicians need to be technically confident and capable to install these stems into their own music sequencer software, record their own parts to a professional standard and upload them back into Dropbox or WeTransfer for the producers.

Working with session musicians is therefore a technical as well as creative process, though creativity should always be a priority and at the heart of the process.

3.5 CREATIVE POP MUSIC VOCAL SESSIONS

It is almost impossible to explain the importance of the vocal sessions in the pop music genre because vocal recording sessions take the producer, engineer and vocalist into a level of higher consciousness that is beyond our daily levels of consciousness. When we are in our daily comfort zones of working, playing and general routines – this is one level of consciousness that we could call ordinary life. Creative actions, such as writing songs, performing songs, producing songs, take us to another level or zone. A vocal recording session, in my view, takes us creatives to a deeper level of concentration, a higher level of consciousness that we also reach at the point where, as creative technicians, we feel we are close to completing a mix. Andrew Scheps (2018) describes this as getting to the point where 'nothing bothers or annoys me' in a mix and he can conclude that the mix is finished. For me, this is a point of creative elation and excitement that you hope the listener will also feel when they listen to the vocals in your mix. Full examples of vocal sessions, with artists such as East 17 at the P&E Music Studios during the 1990s can be found in Chapter 5.

3.6 CREATIVITY IN POP MUSIC MIXING

For me, as a specialist in mixing pop music since the early 1980s, mixing is the most creative part of the pop music production process. Generally, the mixing process is a solo activity so to be a successful mixing engineer/producer with a long career you have to be happy with your own company. Developing strategies to cope with long, isolated mixing sessions is a good idea to retain a creative *flow*. The most obvious strategy is to take breaks at least every couple of hours to rest the eyes from the computer screens and the ears from the music. I have been doing this since working with SSL consoles at PWL Studios in the 1980s, which was before working with Atari ST computer screens.

Be sure to step out of the room and take the opportunity to talk to other people, get some fresh air and perhaps some exercise. One of my strategies during long mixing sessions in the 1990s in Strongroom Studio 2 was to have the control room television on but silent; that made me feel connected to the outside world and less isolated. These days in our multi-screen, small, home mixing studios it might seem a step too far to install an extra screen for watching TV but it could become a tool to keep the creative juices flowing.

There is a full chapter in this book dedicated to mixing (Chapter 6) with a complete run-through of my current mixing techniques that developed from working with so many pop and BoyBands in the 1990s. Vocals were the priority for everyone surrounding those projects: the band, management, A&R and vitally, the fans. Hence, my mixing technique is called 'Top-Down' i.e. vocals first in the mix flow. This keeps the mixer focused on the song at all stages of the mix as opposed to starting with drums and spending hours on them with no song reference present in the working audio balance. So, in my view, starting pop mixes with the vocals will produce the most creative results possible for your clients.

3.7 CREATIVITY IN POP MUSIC MASTERING

Hodgson and Hepworth-Sawyer (2019, p.270) state that 'mastering is a wholly creative process' and to those of us who have experienced the process of comparing our pre-mastered mixes to our post-mastered mixes it is astonishing that some would choose to avoid this process with someone with a fresh set of ears outside of the pop music production team. The mystery of mastering has long been regarded as one of the most difficult subjects to teach in music education. Many college and university courses around the world will shy away from the subject. Most will state that this is a job easily performed by any producer, recording engineer or mix engineer. Experienced industry practitioners, such as producers and mix engineers, will always recommend that involving creative mastering engineers as the end-stop of the production chain is good practice. It is a final opportunity to have your work auditioned and constructively judged by another set of professional ears, trained in the art of critical listening. The 'aesthetic priorities' (Hodgson & Hepworth-Sawyer, 2019) of the mastering process are aligned to what the pop producer and pop mix engineer are also seeking from the product and to not make use of this process in the creative chain is a missed opportunity.

UK mastering engineer, Mandy Parnell, explains the creative critical listening process of a mastering engineer:

> 'I'm listening to what's going on emotionally, the flow and continuity [of an album or EP], and I'd make notes along the way....You need to think about the art first, then the box [technology].
>
> (Parnell, 2019, p.179–180)

Mandy demonstrates the creative aspect of those fresh, professional ears, on the first listen to a project, focused entirely on the art and not the technology. One of the wonderful aspects of the mastering engineer's position is that they have no idea of the background of the potentially difficult process and angst that the producer and artist may have gone through to write, record and present the mix to that final stage, as Barry Grint points out:

> 'We don't know about any of the history [of the project or track]. We just sit there and ask ourselves "how does it sound?"…we only ever work in one room, so you become very familiar with how you expect things to sound in your setup'.
>
> (Grint, 2019, p.167–168)

I don't speak about mastering within this book as I do not consider myself a mastering engineer. My experience with the process was somewhat limited during my busiest and most successful periods throughout my pop music production career (late 1980s–late 1990s). During such busy periods, one rarely has the time to attend the mastering sessions and pop producers will hope and trust that whoever the record label has chosen for the job will deliver a professional and creative result, without over-processing your final mix master from the point of delivery. During that time, I would always recommend who I would like to master my mixes if asked. My first choice was always Richard Dowling at Transformation (now Wavmastering.com) because Richard became the digital mastering engineer at the PWL-owned mastering room in the late 1980s and we got to know each other's work very well. I would refer readers to the excellent mastering book in this series *'Audio Mastering: The Artists'* (edited by Jay Hodgson and Russ Hepworth-Sawyer, 2019).

NOTES

1. ANT stands for 'actor-network theory' which is a tool 'to explain the network through which recording technology is produced and disseminated and the network in which to make music' it is 'notions like power, roles, persuasion and trust' (Zagorski-Thomas, 2014, p.92). There is a collective activity in ANT and again this theoretical approach is captured very well in my discussions around a service model for (pop music) creativity and commerce, see Chapter 8.
2. SCOT stands for 'social construction of technology' and is another systems model that takes a constructionist approach to looking at technology. The cultural domain within the SCOT model sets the boundaries of how we use our technology and interact with each other, the action script, to achieve a creative and commercial result.

4

Pop Music Songwriting

4.1 INTRODUCTION – INCEPTIONS

'At the heart of pop music songwriting and pop production lays a common goal: Popular success' (Hennion, 1983). Hennion talks about musicologists avoiding the subject of economic success and here we are in 2019 with few examples in academia to call upon for the answers. Economic, sociological and cultural studies have progressed since Hennion's theories but in terms of the creators of successful pop music, it is still a paradox. Some might say, as Hennion did, that successful pop songwriters, producers and entrepreneurs over recent decades have a sixth sense, a 'gut feeling' or a 'nose for a hit'. I will highlight two characters, Pete Waterman and Tom Watkins throughout this chapter, whom I believe had this sixth sense. This requires a passion, enthusiasm, fixation or even an obsession for all things musical, especially pop music and a dedication to being successful in popular music. Some might call PWL a good example of an oligopoly, where a small company and group of people prove that you can create an economic condition in which there are so few suppliers of a product that one supplier's actions can have a significant impact on prices and on its competitors. The major record labels of the 1980s were upset and bemused over how a small team of creative people in South East London (PWL & SAW) could take a 40% share of the UK singles market from under their noses. In this chapter I will reflected on the evidence that these entrepreneurs also have a role at the songwriting stage of the manufactured pop and BoyBand phenomena of the 1990s.

I will refer to some of the songwriting techniques adopted by the P&E Music team throughout the 1990s with manufactured pop and BoyBands and will contextualize those into today's practices.

4.2 WESTERN CONTEMPORARY POP MUSIC SONGWRITING

Songwriting is a key component of contemporary popular music. The starting point for the creation of a song might be the musical structure or structured lyrics that will be used to develop musical material. Songwriters, Elton John and Bernie Taupin, have followed the latter methodology throughout their very successful partnership since the 1970s. Simon Frith (1996) states that market forces have affected the pop song 'formula' since the modern song business began in the 1830s. I am going to set out a schematic pop song formula framework that has been ethnographically reflected from my P&E Music 1990s model and updated for 2019. This framework is supported by Case Study #1, referred to at the end of this chapter, and after further analysis led to the theory 'Group Creativity and Human Interaction' in Chapter 8. The main focus of this chapter will be on songwriting, which is central not only to my work but also to the output of BoyBands that form the main discussion. Songwriting in this context has a number of constraints but is always focused on the hit single.

In their book, *The Manual: How to Have a Number One the Easy Way*, Cauty and Drummond (1988) as pop act, The KLF, presented the idea of a formula, which, at the time of writing, would guarantee a #1 hit single. Their idea was that if you follow a specific set of instructions, you have the potential to create a hit record. My own approach to presenting the framework is a more reflective approach drawing on my experience of the last 35 years. The generic approach to writing a #1 presented by Cauty and Drummond omits one of the vital components of songwriting, which is the importance of human interaction within the production team. Although Cauty and Drummond appear scathing about the SAW songwriting and production team, their comments highlight a further important factor in this formula: The deep and genuine love of contemporary pop:

> They [Stock, Aitken and Waterman] are ridiculed by much of the media and only have their royalty statements for comfort. Waterman might be a loud mouthed, arrogant, narrow-minded self-publicist, but the man has never outgrown his true, deep and genuine love of 'Now' pop music.
> (Cauty & Drummond, 1988, p.36: 1998, p.68)

Rather than using the KLF zenarchistic approach to this framework system, this chapter employs a theoretical and practical method. There is a tried and tested pop song formula built around variations on intro, verse, bridge and chorus that has been with us since the 1950s and has modified and developed due to the demands of the contemporary listener and media, which have demonstrated a lack of patience and listening times across all genres of pop music. Therefore the song arrangement needs to be precise and dynamic. Guy Fletcher, former Chairman of the UK performance collection society PRS, recently suggested to me that in his view, the most important aspect of writing a successful pop hit is for it to contain *dynamics*. Pop songs contain three sets of dynamics: Firstly the chord sequence, secondly the lyric and thirdly, in my view, the production. In this way the dynamics of production and the songwriting can be understood as an inseparable process in manufactured pop and it has been recognized that contemporary pop music songwriters have an umbilical link to the production process. Hence we see the songwriting and production credits for many modern pop hits containing long lists of names because people are working in teams to achieve success. Matching skills with other creative people has become vital to the process of apparent solo success for artists, such as Adele, whose songwriting and production teams are in the background. My own experience of creative teams around pop artists and BoyBands has led me to believe that this system, as opposed to songwriters and producers working alone, will stand a better chance of longevity.

4.3 POP AND BOYBAND SONGWRITING: IS THERE A REPEATABLE FORMULA?

P&E Music (Ian Curnow and I) secured a music publishing deal with Mike Sefton and Paul Curran at BMG Music in early 1992. It had become standard practice by the early 1990s that producers were either the main songwriting source of many pop and dance songs or often collaborators with the artists. A healthy market for pop songwriting and production teams had built up throughout the 1980s and this had become a rich source of income for music publishing companies. The publishers often had little to do in placing songs with artists to secure 'covers' as the producers were often collaborating with the acts and assisting with licensing deals to record companies.

P&E Music also agreed a management deal with Tom Watkins in 1992 shortly after the BMG Music publishing deal. No contracts ever existed between Tom Watkins and P&E Music, just a gentlemen's handshake and a commission agreement that started low but rose to the industry standard of 20% by late 1993. It took that trigger point with Tom Watkins before P&E Music began to have some breakthrough success in 1994 with our pop and BoyBand songwriting. Significantly in 1994, Rob Kean would become our P&E co-producer and co-songwriter for the remainder of our period with Tom Watkins and Massive Management. Our careers turned around by the spring of 1994 from 'in-demand' producers and remixers to successful co-songwriters, regularly entering the UK charts in 1995. The period from the middle of 1994 through to late 1995 became a creatively successful time for P&E, much the same as 1987 had been for SAW and PWL (Harding, 2010) in terms of a 'peak creative period' for a production and songwriting team, achieving what Csikszentmihalyi (1992) refers to as 'peak flow'.

A creative peak, drawn from a number of combined factors, is a dynamic and changing collaborative experience. Critical factors in collaborative relationships emerge once initial commercial success has been achieved. An example of this is a situation with East 17 where the formative band-member, Tony Mortimer, had written all the songs (albeit with some collaboration from other professional songwriters, including ourselves on album two) on both of the first two East 17 albums. This led to a situation where the other members of the band insisted on an equal share of the songwriting on the next album once they realized the monetary worth of the songwriting gain to Tony by the fact that he could afford to purchase a much larger property than they were able to. The simple answer to this problem could have been solved in the early days of the formation of East 17 and their collaborative business arrangements. Here is Ian Curnow's view on that period:

> In East 17's case, Tony did all the songwriting up to the third album and in conversations I had with Tony during breaks out-side our studio at The Strongroom, he would say to me 'why does everyone [in the band] hate me?' My answer would be 'well Tony, HELLO, you're doing all the writing [and not sharing it with the other band members]'.
> **Phil:** Do you think that Tom Watkins, with all his management experience with Bros and Pet Shop Boys, should have had the foresight to say to Tony Mortimer that he should be sharing the songwriting credits and splits with the other band members?
> **Ian:** Yes – at least a split that was something like 40% to Tony and 20% to each of the other band members would have seen band survive longer. Bands like Five wouldn't have existed if it weren't for East 17 going before them. That sort of thing is still business but on a more personal level and the difficulty is that you're deal-ing with people and it's a people business – the music industry.
> (Curnow, 2014, personal interview)

Creative collaboration is very much a 'people' business and what Curnow highlights is the vital importance of being fair and honest with the people you are collaborating with. Creativity within a collaborative group activity, such as songwriting, forming pop bands and the production process, is full of pitfalls. Ian's comments are not meant to belittle Tony Mortimer's songwriting qualities and abilities, which are highlighted here by the East 17 session backing vocalist Tee Green:

> Tony Mortimer was a super-talented songwriter, I don't know if he read the bible 50 times or what but his lyrics were incredible. He's had to work hard to do other things within the band but what came naturally to him was lyrics; 'I'll butter the toast if you lick the knife' [from the East 17 song 'Deep'] – I don't know where you get that from to be fair, I have no idea. That wouldn't be in my

head at any day of the week for a million quid! Rather than being out at the Epping Forest country club every night – like the other boys in the band – Tony would be indoors writing those types of lyrics and melody lines.

(Green, 2015, personal interview)

Clearly for Tee Green, Tony Mortimer's lyrics were the highlight of his songwriting talents. Tony would regularly collaborate with other songwriters, generally musicians to take care of the music to support his lyrical ideas. I think we can conclude from this that a BoyBand member and their internal 'band-leader' does not wish to in any way allow conflict with management and record label executives, who are basically running their careers, refusing songwriting collaboration requests in the 1990s and still today would be seen as disruptive behavior within the nucleus that is the manufactured pop and BoyBand genre. Though they may not realize it, a 'band-leader', such as Tony Mortimer, is often accepting these situations on behalf of the other band-members to keep their careers and success moving forwards, even though the other band-members are probably unaware of the sacrifices that someone like Tony Mortimer is making on their behalf. This includes no prior or post discussion between Tony and the rest of East 17 about the songwriting scenarios described by him. Tom Watkins instigated the co-writing situations that Tony Mortimer was pushed into, such as writing with Rob Kean to help Rob obtain a publishing deal. Rob Kean had signed with Tom's publishing company, Porky Publishing, and to obtain a publishing advance from Polygram Music (Porky Publishing's administrator) required these types of songwriting collaborations and commercial releases.

As previously discussed in Chapter 1, Tom Watkins and Rob Kean formulated a new pop band for Massive Management in early 1994 called Deuce and by writing a few songs in advance, in a very ABBA/ dance pop style, this allowed Ian and I to participate in the vocal auditions. These took place at our P&E Music Studio where we would record the vocalists onto the backing tracks we had written for Deuce. Tom and Rob were good at finding the right look and the styling of a 1990s two-boy/two-girl pop band with a mission to be shamelessly 'cheesy' and bubble-gum pop to appeal to pre-teens of both sexes but hopefully the songs and productions would appeal to teenage and post-teenage youngsters, as well as dance floor fans and the gay club scene. This was the first time that P&E were commissioned by a major record company (London Records) to not only produce but also write every song on an album. The initial songwriting and production with Deuce was driven by Watkins, including the direction for the songs and the style of the production, which was early 1990s uplifting dance-pop, always up-tempo for the initial singles, this allowed other remixers to be engaged for further club mixes to achieve maximum promotion for each single. The first single from the Deuce album was

'Call It Love' and the hope was for this to break into the UK top 40 pop charts and the club charts to get an initial groundswell of support for the band with new fans and the media. There was an early discussion with Watkins and Tracy Bennett at London Records to put Deuce forward for the UK entry into Eurovision 1995.

The Deuce singles did very well throughout 1995[1] but the album's performance was disappointing by comparison. East 17's success continued throughout 1995 as well but after the 'fractious' third album, Ian and I decided that we could no longer work with Tom Watkins and Massive Management. This allowed us to concentrate further on our songwriting careers and we began to reach out to some of the BMG Music international offices to seek further co-songwriting collaborations and opportunities. Consequently we would begin to write songs and produce for a number of European BoyBands, such as Caught In The Act (CITA), signed to ZYX Records in Germany, and Get Ready, signed to Virgin Records in Belgium.

On reflection, it is clear to me now that P&E Music had achieved 'creative peak' by mid-1995 and although we would achieve further commercial success in the following years, the successful collaborations with a 'team-leader' such as Tom Watkins would be difficult to recreate. The psychology of the working relationships with Massive Management and their artists had worn Ian and I down in a mere three years since P&E Music started collaborating with Watkins but human nature pushed us to put this behind us and seek new opportunities and challenges.

Having considered creative collaboration within an industry-driven framework, the next section looks at the creative framework for pop songwriting collaborations for pop acts and BoyBands.

4.4 THE 1990S HARDING AND CURNOW SCHEMATICS FOR POP AND BOYBAND SONGWRITING: CO-WRITING WITH A BAND-MEMBER – 'RUNAWAY' BY 'CAUGHT IN THE ACT' 1997

Ian Curnow and I were still working with CITA in June 1997, recording more tracks for their 1998 'We Belong Together' album, plus B-sides for singles that were either the band's own songs or co-writing collaborations between Ian and I with one or two of the band members. Generally, this was in the form of our supplying a backing track for the band members to write lyrics and a top-line melody. We would always include a guide melody with our preferred choice of phrasing for them to work to. This way of writing with a top-line songwriter is generally safer than allowing them to find their own melody (together with its timbre and phrasing). Ian Curnow always viewed the vocal melody as another instrument within the musical arrangement, which was largely his creation, my role was to start Ian in the correct musical direction and framework for the project and then to help with the tempo/arrangement and groove. Once the top-liner came back with the lyrics and melody (hopefully based on ours), we would then help to finalize and contribute to the lyrics. Ian and I were never good initiating the lyrical concept but once that was presented to us, and assuming we liked it, we could help to shape the final lyrical and top-line structure. This approach to songwriting and collaboration worked for us time and time again, especially in the 1990s BoyBand genre. Here is an outline of the schedule for this style of songwriting, which adds up to 5–6 days minimum.

The BoyBand Collaborative Songwriting Timetable

1. Analysis of song arrangement, tempo and key: ½ day.
2. Initial keyboard and drum programing with Ian Curnow: 1–2 days.
3. Record guide lead and all backing vocals with session vocalist: ½ day.
4. Editing guide and backing vocals: ½ day.
5. Record lead vocals with the artist: ½ day.
6. Editing lead vocals: ½ day.
7. Record and edit backing vocals with other band members: ½ day.
8. Final music and rhythm programing with Ian Curnow: 1–2 days.
9. Mix session with Phil Harding: 1–2 days.

This approach to songwriting worked for the manufactured pop and BoyBands of the 1990s. Having a songwriting framework raises a number of questions around the idea of repeatability and sustainability of such a model in terms of contemporary songwriting and production. One of my fundamental questions has been to discover if there is formula for a schematic system in successful pop songwriting. I have been testing the songwriting formula framework recently on some academic sessions with interesting results[2].

4.5 PHIL HARDING POP SONG AND PRODUCTION METHODOLOGY 2019

The production team: There are many ways to approach this system, but the experience of P&E demonstrates that it takes a number of specific skill sets to achieve a result that is worthy of chart success. My suggestion is to create a team to achieve the best result. My typical creative team would comprise of: A team-leader, a keyboard player/programer, a rhythm programer and a lyricist/vocalist ('top-liner').

Direction: I would have someone (team-leader) in the team choose the style and genre of the song, this will involve: Tempo, timbre, rhythm, instruments and sounds. I would choose one or two 'reference' tracks to work from or 'bullseye/plot' as we call them in the industry.

Melody and Top Line: Should the melody be the same direction as the reference track or a different reference? Different is always better in my view.

Lyrics and Syllables: Do the lyric and syllable choices want to be restricted to the melody and top line reference or can they flow more freely? If you allow freedom to move away from the reference, will this lower the chances of the song's potential of being as successful and as commercial as the reference? Yes in my view.

Genre: Does one stay with the same musical genre as the references or does one deliberately move away? Best to move away from the melody reference and adapt that into the direction reference.

Song Arrangement: Does one stay with the song arrangement of the one or two references or does one implant their own tried and tested song arrangement formula? Plant your own.

KLF's golden rule from their *Manual* (Cauty & Drummond, 1988) was that a song arrangement is a 'framework' you can slot the component parts into.

4.6 PHIL HARDING SONG ARRANGEMENT METHODOLOGY EXAMPLE 2019

Intro 1: 4 bars – music/atmosphere/build-up.

Intro 2: 4 bars – relating to the chorus/with some edited vocals ideally.

Optional Link 1: 2 bars – relating to the secondary hook of the song if possible.

Verse 1: 8 bars.

Pre-Chorus or Bridge 1: 4 bars, 'don't bore us, get to the chorus'.

Chorus 1: 8 bars.

Extra Hook Chorus 1: 4 or 8 bars – related to main chorus but catchier? This hook is the part you are going to remember.

Link 2: 4 bars – instrumental or third vocal hook, again related to the main chorus hook.

Verse 2: 8 bars – similar to verse 1 but with different lyrics.

Pre-Chorus or Bridge 2: 4 bars – repeat of bridge 1.

Chorus 2: 8 bars – repeat of chorus 1.

Extra Hook Chorus 2: 4 or 8 bars – repeat of the first extra hook chorus.

Solo or Middle 8: 8 bars – this should contain a musical difference to the verse, bridge and chorus.

Pre-Chorus or Bridge 3: 4 bars – repeat as before.

Chorus 3: 8 bars – repeat of chorus 1.

Extra Hook Chorus 3: 4 or 8 bars – same as above.

Outro Chorus: 8 bars – to stop ending or fade over 16 bars.

The main difference between this and the 1990s pop song arrangement is the extension of the choruses. My research of more recent, successful pop songs, such as 'Story Of My Life' by One Direction in 2014 has led to the this arrangement. We now see much less emphasis today on the importance of the intro and the concept of a 'chorus snippet' before verse 1 has disappeared from most current pop hits. In the 1990s, the first two choruses would be just 8 bars without the 'extra hook' section that has become commonplace now. With reference to the Cauty & Drummond (1988) KLF formula, they call the section that I name *links* as 'hanging bits' and they call my *bridges* 'the bit that takes you from the verse bass riff into the celebration chorus'. KLF consider key changes as something that normally only add dramatic effect at the end of a song that is beginning to flag. Three more KLF golden rules are that a song should have a dance groove, be no longer than three minutes and thirty seconds and have a standard pop arrangement of intro – verse – chorus – 2nd verse – 2nd chorus – breakdown – double chorus – outro. They also say you will need some lyrics but not many.

4.7 COMPOSITIONAL TECHNIQUES

The children's game hangman is an unusual way to write songs but I experienced this with East 17's manager Tom Watkins and it is an accurate way to plot a new song from a previous hit song that one may be influenced by. This allows you to be close to your 'bullseye' to create a hit song for a new artist. The important thing with lyric writing in this system is the syllables relating to the timbre of the music. Each blank underline in the hangman style relates to a syllable from the original song and a new equivalent is found that relates to the original phrasing and in some cases the top-line melody and shape of the main hook of the original song (see Figure 4.1). Even the chord progressions can sometimes be mimicked. If one chooses a production sound that is nothing like the reference song it helps to take the listener's ear and perception away from the origin of the song plot. Many other production techniques are used to perfect the radio-friendly sound of a BoyBand single, I will come onto those in the Chapter 5.

SONG LYRICS

HANGMAN TECHNIQUE

Figure 4.1 Hangman Sketch.
Illustrated by Vince Canning, 2018. Copyright Phil Harding 2018.

4.8 THE STOCK AITKEN WATERMAN (SAW) SONGWRITING METHODOLOGIES IN THE 1980S

Before we move onto the production section I want to briefly look at the SAW songwriting system, which is discussed in detail in my book *PWL From The Factory Floor* (Harding, 2010). In his songwriting blog, Joe Bennett (http://joebennett.net), has these observations to make after interviews with Stock and Aitken:

> Musically, Stock and Aitken had many overlapping skills (they were both composers, lyricists and producers) but had less interest in the business deals and media networking at which Waterman was so adept. Creatively, the Stock/Aitken dyad generated songs consistently according to an agreed co-writing methodology. The arguably 'non-creative' Waterman contributed not only business opportunities (before and after the co-write session) but also creative constraints (via the artists that he selected) to which the core dyad was able to respond. Despite Waterman's lack of musical or literary contributions, he could arguably still be considered a collaborator – because without his involvement these songs would not only have sounded different, but would have been less likely to become hits. He introduced business constraints – singers, deadlines, target markets, clients – that materially influenced the musical and lyric content of Stock and Aitken's work. We can therefore argue that Waterman was partly responsible for creating *value* in the song, even though he did not write any part of it.
>
> (Bennett, 2015. Available at: http://joebennett.
> net/2015/02/27/saw-kylie-lucky/#more-4305
> (accessed 20 March 2016))

It is clear to me that Bennett did not speak to Pete Waterman to obtain a more triangular view as Waterman and Stock have agreed in the their books that he contributed to the song title 'I Should Be So Lucky' by Kylie Minogue[3]. In the early days at PWL Studios I witnessed Waterman contributing to the songwriting titles and the start point of the song subjects many times. These starting points, however small, are important to the process. Bennett is absolutely correct to say that Waterman would create *value* in the song but it is also his contributions as the 'person with novelty in the cultural and creative domain' (Csikszentmihalyi, 1996) that are useful in this type of creative collaboration. Waterman's introduction of constraints 'singers, deadlines, target markets, clients' to the field, as Bennett points out, would usually have a positive effect. Once a project such as the song 'I Should Be So Lucky' by Kylie Minogue was completed, Waterman would then step back in to generate the commerce for the team out of the creative result. He pitched the track to many of the major record companies in the UK at that time (late 1987), with no success, so Waterman released the single on his own PWL Records label and assisted in steering the record to #1 in the UK charts in January 1988. For the 1990s, I would place Tom Watkins in the same team-leader/ entrepreneurial position as I describe for Pete Waterman. The Watkins creative input to the songwriting and production dyad that was Ian Curnow and myself was invaluable. Whilst he sometimes shunned songwriting credits for diplomatic and political reasons, his studio contributions far surpassed the studio time that I saw Pete Waterman contributing to the SAW projects of the 1980s.

4.9 SONGWRITING FINAL NOTES

Various reference methods (or 'bullseyes' and 'plots' as I have noted) have been used in professional songwriting and production teams for many years and in general terms, many pop songwriters and producers say, 'we have allowed ourselves to be influenced by'. The communities around 'Tin Pan Alley' in London during the 1960s and the New York City Brill Building (Carol King, Gerry Goffin and Neil Sedaka started their songwriting careers there) are examples of communities encouraged to work as teams to write pop songs for chart artists in whatever the current style or pop formula may be in the hit parade by other artists at the time.

Example characters that possess the elusive talent; 'a nose for a hit' could range from Pete Waterman, Nicky Chinn, Mike Chapman and Mickie Most through to Tom Watkins, Simon Cowell and Louis Walsh. The list could be endless and I have touched on some of the British examples of successful producers, songwriters or entrepreneurs that I've known, admired or worked with. Tom Watkins is an example of the entrepreneurial manager who is so obsessed with having control of the project from the beginning (sitting in on the songwriting sessions and writing the lyrics), through to approving the final mix sessions in the studio that he will then take complete control over the cultural and economical aspects of the act. From design, image, promotion and record plugging through to controlling the royalty statements, contracts and every aspect of mass popular success with his artists, ranging from the Pet Shop Boys and Bros through to East 17. All of this shows that some people will stop at nothing to achieve their goal of pop music success. Pete Waterman set out in the 1980s to create a UK equivalent to Berry Gordy's Motown Empire of the 1960s, even calling his company The PWL Empire so that no one was mistaken as to what he was setting out to achieve.

4.10 SONGWRITING CASE STUDY #1

'Is There a Music Composition Formula to Create Successful Pop and BoyBands?'

This question drew a number of differing views from my interview respondents. There were a number of resounding 'Yes' answers and an equal number of 'No' answers. I decided to test an updated version of the P&E Music formula from the 1990s in an educational setting. Across a weekend in February 2015 at The Westerdals School of Arts, Communication and Technology in Oslo, I was presented with 16 students to collaborate with by their course leader Jan-Tore Driesen. Jan and I created four groups of artistically compatible songwriting and production teams based on the individual skills of the students. The results of the task have enabled me to realize that human interaction in this type of collaborative group activity can create unexpected and surprisingly varied results. This research project has developed into my commentary on 'Group Creativity and Human Interaction' in Chapter 8.

Each group contained a 'team-leader' or producer, a keyboard player/programer, a guitarist (or 2nd keyboardist) and a lyricist. I gave each group the same tempo, key and artist to write for and the same reference tracks to work from. The example song arrangement and production ideas were as set out in section 4.5 of this chapter (Phil Harding Pop Song and Production Methodology). We hired a session singer on the second day to perform the lead and backing vocals on the song demos. Each group was given one hour with the singer and I acted as 'executive producer' to assist during the recording. Surprisingly, each team came up with an entirely different song with no similarities to each other whatsoever (the compositions were arranged in separate studios on day one). The results were astonishing to me and basically demonstrated that even though the formula suggested here seems very regimented and even robotic, once human interaction begins in a team, very different results will develop, even in an incredibly short space of time.

The Task: To write and produce (to 2015 music industry song demo standards) a new song for one of my current production artists (Tina Charles), signed to Prolific Music UK.

The Formula Framework

1. The tempo was set at 128 bpm.
2. The key was set to suit the female artist – d major.
3. A recent British BoyBand pop hit was presented as a reference for a modern production style.
4. A hit song with a different famous female singer was presented as a reference song style.
5. The song arrangement from earlier in this chapter was presented to follow as closely as possible.
6. A reference song that my current songwriting team, PJS Productions, had written for the same female artist was also presented as a guide reference for the singer's character but this song was not to be copied.

The owner of the record company kindly offered to listen to the resulting four songs intended for his artist Tina Charles, here is the feedback from Carl M. Cox of Prolific Music UK:

To: Phil Harding
From: Carl M. Cox
19/03/2015

Re: Student songs
All songs are remarkably good compositions and I have really enjoyed listening to them. It goes without saying that each group have really met the brief you had set to write a suitable song for Tina.

'Stand Up'
Any song with a succession of 'woh-ohs' in the chorus is always guaranteed to strike a chord with dance audiences. That is invariably the hook on this number. The song encapsulates the anthem sound to a tee and would work well as a high octane-fueled dance track. For me, in a real-world situation, this is the song least likely to work as a Tina track.

The repetition of the 'stand up, stand up' line sung in counter with the 'woh-oh' makes for a very infectious and memorable chorus, essential for any song. So, this group has made sure that those base elements are there.

The kick drum on the production kind of lets the production down as it is too weak for such a power driven type song and isn't really the kind of pad sound you'd expect to have on as track of this nature. The minimalist verse serves to give more impact on the chorus and this is a commercially very sound technique to use. The breakdown chorus with the hand claps works to great effect here also.

'You Make Me Whole Again'
A great soul infused pop song. It's catchy and has a nice hook with the refrain between the end of the chorus start of the verse. I could imagine Tina recording this song and it sitting well on her album. I would be very happy to consider it for Tina's album. There's a song on the album she's recorded as a duet that I would rather not use. There could well be a space for this song. The song itself is nicely balanced. The stripped back verse building up to the chorus works tremendously well and is another nice hook. The kick drum and toms are another let down on the production. But that is the only issue really.

'Free Your Mind'
This is another great dance inspired track. The chorus on this however, is lacking somewhat. It's not as striking as the other three, but then again, that could be in part that I have had longer to get used to the other three.

Of all the songs, the guys who have written this have created a more standard intro. It has the best intro of all four songs. The others have quite sudden, even abrupt intros. This song has a great structure too. From a commercial viability perspective, the foundations are certainly there. [This song was supplied some weeks after the other three due to production delays].

'I Can't Wait'

Of the four songs this is by far my favourite! It's brilliant and literately leaps from the speakers the moment you play it. From that point it's as instant as 'Bisto' and hooks you right into it! I've had this on multiple repeats and still haven't got bored of it. The verses are pulled back a little, with a hint of the intensity that awaits in the chorus. The chorus then echoes that same euphoria that 'Better The Devil You Know' does. To me it's the audio equivalent of an explosion! An explosion of sounds!

This song oozes commercial potential! Either as a dance track or with a re-arrangement to sound more like Megan Trainor's recent songs. Phil, I would really like to publish this song and record it with Jade [another Prolific Music artist]. It's perfect for her. What would I need to do or whom would I need to speak to in order to progress things? We are recording a second track with Carl and Jade on Monday and if I can reach an agreement with the writers, Jade will record this! That's how smitten I am with it.

(Cox, 2015, personal correspondence)

Some enlightening music industry feedback from Carl Cox of Prolific Music on the four compositions, this kind of feedback is useful in academic circles (and thus this book) and is being used increasingly to allow students to gain insights from industry practitioners on how they view the student's creativity in terms of commercial market standards. I did not expect Carl Cox to like one of the songs so much that he would consider recording it with Tina Charles or another of his artists. The four songs can be found at: https://soundcloud.com/user-606698363

NOTES

1. UK single chart positions: 'Call It Love' #11, 'I Need You' #10, 'On The Bible' #13.
2. See Case Study #1 in this chapter.
3. Matt Aitken has denied this on the recent Sky Arts program 'Trailblazers', broadcast 19 June 2016.

5

Pop and BoyBand Production

5.1 INTRODUCTION: BACKGROUNDS, PRODUCE ROLES AND CONCEPTS

There are a number of roles for producers that I will highlight with commentary from producers, Steve Levine and Richard James Burgess. They have confirmed my long-held views that a vital part of becoming a successful producer, or production team, is to have the ability adapt to any situation that may occur in a recording studio. Therefore, I will highlighted the 'ideal' scenarios and conditions that producers will surround themselves with to cope with vocal recording sessions in order to achieve a satisfactory creative *flow* that will encourage the artists, after recording their vocals, to leave the studio feeling that they have achieved and participated successfully, and importantly, to pass that message back to their representative management and record companies. Mihaly Csikszentmihalyi (2002) states that music is organized auditory information and that when seriously attended to as a listener can induce *flow* experiences. I would say that music producers are fully immersed in the *flow* of music throughout their projects and are good examples of an autotelic personality at a deep level. I will supply detailed information in this chapter of the P&E Music production workflow throughout our 1990s pop and BoyBand projects. These are good examples of a production team's commitment to the goal of servicing the creative needs of the artist through to the commercial requirements of the manager and record company.

Artists and Producers

I had never considered what Negus (1992) calls a 'marriage between artist and producer' to be the relationship I was entering into but on reflection, he is startlingly accurate. I will highlight throughout this chapter the types of working relationships that producers have in this music genre, stating that the only time artists and producers are together during the recording process is for the vocal sessions, unless there is also a songwriting collaboration in process. Nevertheless, the vocal sessions for many different music genre recordings can be the most delicate part of the process; setting the right studio atmosphere, being sure that the technology is working efficiently so as not to disrupt creative *flow*, encouraging the artist constantly regardless of tuning and timing errors, and so on. It is a constant hand-holding, confidence-inducing, ego-massaging process where the producer's self-consciousness will naturally disappear in order to 'service' the artist in the cause of capturing the best vocal performance possible for the song. For the producer, and probably the artist, the sense of time becomes distorted. This closely matches Csikszentmihalyi's (2002) 'conditions of *flow*'. With the assistance of my interview respondent and session backing vocalist, Tee Green, we capture an interesting overview of the life and culture surrounding the 1990s BoyBand phenomenon in Chapters 1, 2 and 7, with commentary covering touring and life outside the music studio.

5.2 MUSIC PRODUCER ROLES

The role of the pop and BoyBand producer is intrinsically linked to the songwriting process. Often, the producer will be involved in the songwriting, either for the artists or sometimes with the artist, as I have outlined in the previous section. There can be a thin dividing line between pop producers re-arranging a song and where that may tip over into warranting a songwriting credit. What is clear from my many years' experience as an industry practitioner in this genre, is that unless this collaborative possibility is negotiated in advance of the recording sessions, it will be difficult for a producer to make a case for a songwriting contribution. Presenting a co-songwriting negotiation claim once the music production is complete will prove difficult and likely lead to a breakdown in relationships between the producers and their clients, whether that is the record company, management or the artists themselves. One risks not only the abandonment of the current project but it also jeopardizes future projects from that client team. The role of a music producer has been widely discussed in the media and literature. The BBC Four television program 'Music Moguls – The Producers' (BBC, 2016) allowed a number of producers to air their views on production roles that can vary from being a catalyst that stimulates ideas[1], even though they may have little musical knowledge themselves, through to facilitators that provide recording studios and commercial outlets for artists to express themselves. The facilitator type of producer will often exist in the rock, jazz and classical music genres. In the program, Steve Levine comments on 1960s pop producer and songwriter Phil Spector suggesting that:

> One of the other roles of the record producer is to find talent and Phil Spector, to have discovered the number of artists that he did, that's quite a skill. He didn't do it once, he did it several times,

that is the role of a producer, not just making records, it's putting together the artists and the musicians to get that *thing* that really works. He was able to find the songs and he had the knowledge and understanding of what was a great song.

(Levine, 2016, BBC Four TV)

Phil Spector's multi-layered, overdubbed 1960s 'wall of sound' was unique at that time and he called it his 'Wagnerian approach to Rock n Roll'. George Martin's production style with The Beatles started out as facilitator but once the band members began to embrace Martin's orchestration and arrangement talents it became more of a musical collaboration as their relationship developed. Hence, Sir George Martin will always be lovingly known as 'The Fifth Beatle'. Within the same program, producer and arranger Tony Visconti states 'there are no rules [in music production], you can do what the hell you want, that's what George Martin taught me'. It can be said that no two producers or production teams will have the same methodologies or identical skill sets. What I present in this study is that we can define what methodologies and skill sets are required to become a successful production team, specifically in this manufactured pop and BoyBand music genre. Richard James Burgess (2013) sets out six categories of music producers that he has derived from observing working practices:

1. Artist: Artists who produce themselves.
2. Auteur: The primary creative force in the production.
3. Facilitative: The artist is the primary creative force.
4. Collaborative: The producer shares the creative load.
5. Enablative: Enabling conditions in which a successful recording could take place.
6. Consultative: A mentor, often not present at the studio.

(Burgess, 2013, pp.9–19)

Burgess supplies an interesting narrative of each category that I have summarized for my purposes here, he also goes on to talk about leadership styles that Daniel Goleman (2000, cited in Burgess, 2013) lists in his book *Leadership That Gets Results*:

1. Visionary: Come with me.
2. Affiliative: People come first.
3. Democratic: What do you think?
4. Coaching: Try this.
5. Pacesetting: Do as I do, now.
6. Commanding: Do what I tell you.

(Goleman, cited in Burgess, 2013, p.24)

These leadership styles could relate to any industry or creative practice, but Burgess relates this to music production: 'The best leadership strategy is to make the artist and label feel confident and engaged in the process' (Burgess, 2013, p.25). Using these two lists allows me to narrow down the definitions for my 'Service Model For (Pop Music) Creativity and Commerce' theory where the team-leader or entrepreneur needs to be like (1) a visionary and a frontrunner (respected by the team) in their role and yet only (6) in their production role. The rest of the production team need to be a little bit of everything Burgess lists from (2–5) in their collaborative production, musical and technical roles. My suggestion is that once all of these skill sets, elements and roles are in place, creativity and commercial *flow* can be achieved in pop and BoyBand production.

1992 was the beginning of the Harding and Curnow period for testing and formulating a new team of collaborators around us to fulfill our ambition to become innovative pop producers and songwriters at the highest possible level in the UK. Our aim was to break free of the PWL stigma and sound of the 1980s and to be the center of a creative, collaborative team that could consistently score UK chart hits with new talent that we would help to discover and develop. We knew from our eight years of experience at PWL that we needed a guiding figure that could be our manager/team-leader. Our choice was Tom Watkins and we had some experience of working with Tom on various remixes and productions in our time at PWL. I knew him to be a powerful, hard-hitting entrepreneur and dealmaker. Just the type of person we needed. Here is Ian Curnow's response when I told him that Tom had recently told me that he thought East 17's 'House of Love' was still our best piece of collaborative work with him:

> I agree with him in many ways because at that time we were like the rabbits in the headlights and we just went for it and I think that's what separates the good from the bad, not wishing to sound pompous. At that moment in time we were on it, we were a hot remix and production team willing to take risks and work with new technology. At those moments, if you dig deep you can achieve something spectacular, which we did with 'House of Love'. We were also fresh out of leaving PWL, keen to prove ourselves and this was a really exciting project for us to get our teeth into. It fired us up – Tom came in with tons of energy and enthusiasm about the project. Give us (P&E) the goalposts and we'll go for it.
>
> (Curnow, 2014, personal interview)

A very succinct view from Ian; Tom's energy and enthusiasm was infectious and inspiring to both of us and it would drive our individual and collaborative roles in the recording studio. We agreed a producer management deal with Tom Watkins and his Massive Management team, hoping that more success would follow East 17's debut hit single. Ian and I produced some of the tracks for East 17's

first album, 'Walthamstow', and by the time we were producing the
second album in the summer of 1994 we had formulated a pop and
BoyBand system for the all-important vocal recording sessions. The
vocal arrangement and recording process that I describe next is a
typical methodology of pop producers, such as Phil Harding and Ian
Curnow, working with pop and BoyBands in the 1990s. This process
would begin once the basic backing track was programed and in the
correct key for the lead singer. This example is from East 17's 1994
'Stay Another Day' single.

5.3 HARDING AND CURNOW 1990S POP AND BOYBAND VOCAL PRODUCTION METHODOLOGY

1. Ian and I would book our favorite session singer (Tee Green) for
   an afternoon and set about mapping out the vocal arrangement
   with the production team musical director (Ian Curnow in this
   case).
2. This is done before the artists come in, so we would require the
   session singer to record a guide lead vocal to enable the analysis
   of the backing vocal arrangement.
3. Firstly, Ian and I would record the chorus harmony parts at the
   first chorus; each part would be tracked (re-recorded) at least
   twice to allow each part to be panned in stereo. In the case of
   'Stay Another Day' we recorded each of Tee's chorus parts four
   times to create a fuller sound. We would then copy those backing vocals to each chorus; there is no need to sing all those parts
   again unless there is a key change, which is often the last chorus.
   These parts would have been recorded to multitrack analogue
   tape, transferred into Cubase Audio using Digidesign hardware
   and copied within Cubase. Usually recordings three and four of
   each part would either be sung softer or harder than takes one
   and two to obtain a difference in sound, this contributes to the
   vocals sounding thicker and fatter.
4. Ian and I would then go back across the verses, bridges and
   middle section to see if any other lines need harmonizing. Often
   just a third harmony above the lead is enough but regularly with
   BoyBands there will be block backing vocal answers in the verses
   and other sections. This is to keep the other band members involved and featured during television and video performances.
5. Having recorded everything the producers might have been planning, one would then give the session singer the opportunity to
   see if they have any further backing vocal ideas that might have
   occurred to them, whilst spending all afternoon singing one
   three- to four-minute pop song. Tee Green always had many extra
   ideas that we were happy to give him time to record. In the case
   of 'Stay Another Day', Tee instigated most of the backing vocal

ideas in the verses. We had one last backing vocal idea for the big-production ending that songwriter Tony Mortimer had given us freedom to experiment with and that was a counter-melody hook to come in from the third repeat chorus of the outro.

6. A day or so later, on completion of editing and copying all of the above, we would bring either the whole band into the studio or preferably the main lead vocalist for the bulk of the day to record the lead vocal throughout the whole song (recording multiple takes to choose from), usually double-tracked on the chorus section. Then either later the same day or maybe even the next day, each BoyBand member chooses, or is recommended, a chorus harmony part and records their version of it twice, singing along, often in solo, with what has been recorded with the session singer. In the case of 'Stay Another Day' we had to find a part for Tony Mortimer to sing, as there was no rap section in the song, so he sang the 'stay now' line that leads us into each chorus. In my 2014 interview with Tony he remembers that Ian and I gave him a chance to sing the whole song. It was clear to him and us that this wasn't going to work but we were happy to achieve the important hook line 'stay now' with Tony.

7. It can then take at least half a day to compile, edit, time and tune all of these vocals described. The software Autotune arrived around 1998, too late to help us with most of the East 17 recordings but Autotune and Melodyne have been a big help in the vocal editing process since. These vocal processing methods are also a big help in the recording process because one can record a good 'creative performance' from a vocalist and one now knows that even if it is slightly out of time or slightly out of tune it can be retuned and re-timed. This allows producers to concentrate on capturing character and emotion in vocal performances.

8. Most BoyBand productions will end up with something like fifty vocal audio tracks out of a total average audio track count, for me, of over one hundred. Modern technology and digital audio workstation programs (DAWs) such as Avid's Pro Tools and Apple's Logic Pro have allowed us to achieve this.

9. The final part of the process occurs during the mix where I would balance the band backing vocals about 5db below the session singer backing vocals, just enough to add some character and fullness.

10. Some people would be surprised when they listen to our typical 1990s pop tunes and we would say that was at least a five-day recording process and my mixing would be at least one and a half days on top of that. As can be seen, the vocal recording and editing process of a typical 1990s BoyBand song would stretch across at least three days.

This system will still work with pop acts and BoyBands today and I am still practicing this methodology with my current pop production team, PJS Productions. One can see the advantage of recording the backing vocals first from my methodology. It allows producers to present an efficient template to the artist to work from. A demonstration recording of East 17's 'Stay Another Day' was recorded at a small London studio by Dominic Hawken and delivered to us at the Strongroom Studios on cassette by Tony Mortimer. Once we had confirmed with Tony a song arrangement structure that allowed us the scope to create a piano and orchestra-led track, Ian set about programing this into our Cubase music sequencer.

5.4 ANALYSIS OF THE POP AND BOYBAND PRODUCTION FORMULA FRAMEWORK

Many producers in the rock music genre will take an organic approach to building a production in the studio, taking views from band members, session musicians and the band's manager, allowing the tracks to grow naturally throughout the recording process. By contrast, a pop music producer, especially in the BoyBand or manufactured pop genre, will often have a strong vision of the finished record before stepping into the studio or before the first note is struck on the keyboard for the initial programing of the record. Whilst my framework formulae are strictly structured, they allow for deviation as designated by human interaction, creative ideas from members of the team, and most importantly, going with the *flow* of how the song and production shape up dynamically. There is an outline and commentary of a typical P&E single-track production timetable in the production appendix.

Production Logistics

By 1994, Ian Curnow and I felt we were part of the Tom Watkins 'A-Team' as opposed to the PWL/SAW 'B-Team'. This was one of our motivations for leaving PWL in 1992 and exactly where Ian in particular wanted to be, at the heart of the songwriting team for a new act such as Deuce. For a production team working in this music genre, it is vital to accommodate the artists in a comfortable and welcoming environment. Often the whole band may arrive together, even though the schedule was to record only the lead vocalist on a single song, which could take up to five hours. Ian and I started taking steps to expand our operation to be able to accommodate bands like Deuce and East 17 entering our studio at the Strongroom. We spoke to Strongroom owner, Richard Boote, about extending our production suite into an adjoining space that could include a comfortable lounge with sofas, a 5.1 cinema system, a kitchen and even a small, additional, programing room where artists could experiment with song ideas. Importantly we would have our own direct entrance from the Strongroom car park, which gave the impression to visiting artists and clients that we 'owned' a larger space than any of the other Strongroom project studios that shared entrances.

Post Tom Watkins

By the summer of 1995 it was clear to Ian and me that it was time to move on and regain our independence from Tom Watkins and Massive Management. I am not sure if we felt it at the time but there was a clear déjà vu with what had happened to us in 1991 at PWL, when the production deal for us to produce Take That fell through due to Pete Waterman's decision not to allow the PWL credit to be removed from the contract and therefore the Take That record credits. That triggered a clear reaction from Ian and I that it was time to move on from PWL because it was becoming detrimental to our work and our future careers to stay there. The same happened in 1995. Our best work with Tom Watkins and East 17 was probably behind us and in terms of Tom Watkins 'sticking up for us and fighting our corner' (as one hopes your manager will), it was not happening. It was another 'light bulb' or gut reaction moment for me except unlike in 1991 where we stayed at PWL for almost another year, here we knew it was just a decision to fulfill commitments within two months. Ian and I were confident that once the word spread that we had left Tom Watkins, work and offers would continue for us as a songwriting and production team. We had our publishing deal with BMG Publishing and support from Paul Curran and Mike Sefton, who had been delighted with the run of money-spinning hits we had delivered in 1994 and 1995 with Deuce and East 17.

In 1996, Ian and I signed a new management contract with Zomba Music. Although I talk about other artists in the P&E Studio throughout late 1995, most of 1996 and into 1997 was dominated by Louis Walsh's acts, certainly in terms of P&E chart success with Boyzone and OTT. We were seen by many as Louis Walsh's producers throughout this period and commercially we were happy to be seen so. I realized by late 1996 that Louis Walsh was good fun to work with and I really enjoyed his company. His sense of humor and his flippant attitude to the music industry (we have seen that on the X Factor over the years) made him easy for P&E to work with. He had, and continues to have, a light-hearted and good-humored attitude towards everything – always optimistic and positive. If he did become stressed or worried he never showed it. I will move on now to examine the artists in this music genre.

5.5 PRODUCTION SYSTEM AT P&E MUSIC STUDIOS

A Schematic Approach to the Harding and Curnow Signature Production Sound for 1990s Manufactured Pop and BoyBands

The Production Timetable

1. Analysis of song arrangement, tempo and key: ½ day.
2. Initial keyboard and drum programing with Ian Curnow: 1–2 days.
3. Record guide lead and all backing vocals with session vocalist: ½ day.
4. Editing guide and backing vocals: ½ day.
5. Record lead vocals with the artist: ½ day.
6. Editing lead vocals: ½ day.
7. Record and edit backing vocals with other band members: ½ day.
8. Final music & rhythm programing with Ian Curnow: 1–2 days.
9. Mix session with Phil Harding: 1–2 days.

Figure 5.1 Ian Curnow (left) and Phil Harding (right) at their P&E Music Production Suite within Strongroom Studios, circa 1993.

The example timetable can become up to nine days of work and is similar to the songwriting timetable in Chapter 4. It represents a typical East 17, Boyzone or Deuce production in the 1990s. If Ian and I had been involved with the songwriting then much of the music programing and backing vocals would already exist. There are a number of examples of specific P&E productions and process in Chapter 7, including the popular 'Additional Production and Remix' version that was a common credit for Ian and I throughout the 1980s and 1990s. That credit describes where production teams will replace most of the music backing track around previously recorded vocals from another producer or production team, the 'House of Love' commentary in Chapter 7 is an example of that scenario. The mix process is the last vital stage for the pop and BoyBand music genre production process and my extensive commentary on this can be found in Chapter 6, featuring 'The Phil Harding 12-Step Mixing Programme'.

NOTE

1. Producer Steve Levine uses 1960s pop Svengali, Joe Meek, as an example.

Pop Music Mixing

6.1 MIXING 1990S MANUFACTURED POP AND BOYBANDS

Richard Boote, owner of The Strongroom Studio complex, had seen the potential for the new idea of production suites for producers, remixers, songwriters and artists in the late 1980s/early 1990s and The Strongroom were one of the first major UK studios to provide these spaces for rent. It was in September 1992 that Ian Curnow and I moved into our new, purpose-built, soundproofed studio at The Strongroom. Production suites such as these are now becoming common practice for studios throughout the world. It is a win-win situation for both parties because the studio owners gain rental income and many of the production teams will need to hire the larger studios within the complex for recording drums, bands and overdubs from numbers of musicians that cannot fit into the production suites. P&E would also hire Strongroom studios 1 and 2 for mixing on major projects, using the Neve mixing console in studio 1 or the Euphonix in studio 2. Later in the 1990s, The Strongroom would add studio 3 to the building that housed our production suite. It was equipped with my favorite SSL G-series console, which had been installed into studio 2 prior to the arrival of the Euphonix. All analogue mixing desks add a sound characteristic to the audio signal path and it is difficult to define what one engineer or producer may favor over another. My experience over forty-five years in the audio industry is to align the sound of a mixing console to the genre of music that it most suits. The Neve mixing console has a reputation for producing a warm sound, aligning it, in my view, to acoustic instrument recording and mixing, such as rock, folk, jazz and classical music. The SSL console is renowned for producing a sharp sound, aligning it to being beneficial for pop and dance music. I acknowledge that these are sweeping generalizations but go some way towards informing the technology choices we make as creative technology practitioners.

The beneficial part of hiring the other studios at The Strongroom for mixing was that this allowed Ian and I to double up on our work space, Ian would remain in the P&E studio, programing our next tracks or projects whilst I was in Strongroom 1 or 2, spending one to two days mixing. We would regularly visit each other throughout those days to check progress and comment on our work. If these Strongroom mixing rooms were unavailable due to other bookings then I would regularly book either Trevor Horn's Sarm East Studio or go back to my old mix room, The Bunker, at Pete Waterman's South London studio complex. Both of these studios were equipped with the SSL G-series mixing console. Surprisingly, a mix project for East 17 and London Records would provide proof that we could also mix successful records in our P&E studio production suite, diminishing the need to always use the larger mix rooms available within The Strongroom complex. Initially this was a case of confidence for me because the other studios that I have mentioned had a reputation and proven track record of chart success. For P&E, there was a need to see a hit record in the UK charts that had been mixed in our studio to validate the process to our clients and ourselves.

East 17's third single 'Deep' became a top ten UK hit and provided the evidence that Ian and I required to justify mixing in our room for certain styles of product that would benefit from a slightly less polished sound than was generally produced by the other Strongroom mixing studios. One of the major factors towards this difference is that the SSL and Neve consoles provide mix automation, which allows the mix engineer to automate fader and channel mutes many times, similar to re-recording vocal performances. The Soundtracs analogue console in the P&E Music studio (see Figure 5.1) did not have mix automation, therefore a mix had to be achieved in a 'live' pass, which would include all fader movements and all fader mute requirements. This was only ever possible for us on material that was quite rudimental in its production. Another act that benefitted from this mixing process was J-Pac, a Massive Management act that were signed to Epic Records with a sound that needed to be street-savvy to match the largely rapped and shouted vocal styles combined with very bare backing music.

As the Apple Macintosh computer became faster and more powerful throughout the 1990s, Ian and I became more confident in delivering final mixes completed within our P&E studio. By the late 1990s I would run the 24 Cubase Audio outputs through the Yamaha O2R mixer that we had purchased in 1996, whilst the 24 tracks from the Saturn tape machine returned through the Soundtracs mixer. This developed into a powerful and flexible system. The Jamie Shaw project was one of the few in the late 1990s that would still require Strongroom studio 3 to complete the mixes, which contained fully arranged orchestras (recorded at various studios)[1] and the David Grant choir, recorded in Strongroom studio 1[2].

Since the mid-1990s I have adopted a unique schematic system for my popular music mixing process and it is only in recent times that I have talked about this system publicly and presented the framework of a mixing workflow to production and music technology students. During these presentations I have become aware that not only had the students never heard of this system before but also that many lecturers were unfamiliar with it as well. I have therefore developed the following program; reflectively based on my own mixing practices.

6.2 TOP DOWN MIXING: A 12-STEP MIXING PROGRAM

This system is a reflection of my working practices as a producer, engineer and mixer in the 1990s. In what follows, I suggest, that the methods I used then are still useful in the context of modern popular music mixing. To do this, I conceptualize mixing in two different ways, specifically, as a 'top down' and a 'bottom up' creative practice. 'Top down' in this concept, refers to starting a mix with the lead vocals and then working 'down' through the arrangement to the drums. 'Bottom up' refers to the opposite. 'Bottom up' mixing begins with the drums and ends with the vocals. The latter method has, in my experience, been the traditional routine in rock, pop and dance music genres since the 1970s.

'It will be all right at the mix' and 'it's all in the mix' (Cauty & Drummond, 1988, 1998, p.117) are phrases I have heard in recording studios since I started making records in the mid-1970s. By this time a multitrack recording, requiring a final stereo mixdown session, was already a well-established part of the production process. The mixing process is barely touched upon in *The Manual (How to Have a Number One the Easy Way)* (Cauty & Drummond, 1988, 1998) by Jimmy Cauty and Bill Drummond of The KLF, first published in 1988, and I do not go into the 1980s SAW mixing process in any great detail in my own book, *PWL From The Factory Floor* (Harding, 2010). I would agree with William Moylan (2015) that 'people in the audio industry need to listen to and evaluate sound' and that prolonged periods of critical listening 'can be used to evaluate sound quality and timbre' (Moylan, 2015, p.186). Moylan raises the important question of timbre, which Zagorski-Thomas (2014) notes 'is a function of the nature of the object making the sound as well as the nature of the type of activity' (Zagorski-Thomas, 2014, p.65). Ultimately, the mixing process is something that needs to be performed, practiced and mastered by every creative music person in this time of diversity. Composers, musicians as well as engineers and producers, will find themselves in the mixing seat, due to budget or time constraints in an age where budgets for all types of recorded music and audio have fallen by 25% or more since the 1990s. The following 12-step program still serves me well after forty years of experience as an industry practitioner and is a framework for others to experiment with[3].

It was during my BoyBand production period in the 1990s that I tested the idea of starting a mix with the vocals. Possibly, it was the large production-based projects that Ian Curnow and I did with East 17, and in particular their 1994 Christmas #1 single 'Stay Another Day', that led to my testing this system. I was mixing this track in the summer of 1994. It was the day before going on a family holiday and I had a huge mixing task in front of me: Over 50 vocal audio tracks together with multiple keyboards, drums, bass and a full orchestral arrangement, programed by Ian Curnow, including the Christmas tubular bells. All of this was across 48 tracks of analogue tape so some sub-mixing had already been done in Cubase Audio in our production suite at The Strongroom Studios. Nevertheless, there were at least 10–16 tracks of vocals to be dealt with, as quickly and efficiently as possible, so that I could get home at a reasonable time for our holiday journey the next day. Apart from the usual lead and harmony vocals from vocalist Brian Harvey, there were also counter chorus vocals from band member, Tony Mortimer, and four part chorus harmony vocals, all double-tracked, from each band member in the chorus. Then there were a large quantity of chorus harmonies, each quadruple-tracked by session vocalist, Tee Green, plus verse harmonies and answers from the band members and Tee. Finally, there was a backing vocal counter melody on the outro of the song, again quadruple-tracked by Tee Green. As the family holiday deadline loomed, I had a 'light bulb' moment and decided that I would start the mix with Brian's lead vocal, supported by the main song pad synthesizer for some musical perspective. From that day on, this has been my adopted mixing method for every record I have mixed, regardless of genre. I believe this does not only work for BoyBands or pop music, it can work for any mix that contains vocals. One may need to adapt the starting null-point for other digital audio workstations (DAW) and software but this has worked very well for me on hardware such as the SSL G-series and Pro Tools software. Here is my '12-Step, Top Down' mix routine.[4]

Step 1: Main Lead Vocal and Master Fader

For my first step I would set the main lead vocal fader to zero. That can be the group master or the individual fader if I have more than one lead vocal track (crossovers, and so on). I also set my master fader to zero. Hopefully, if the vocals are well recorded, this will be a good start point and hopefully a good null-point setting for the mix overall.

Step 2: Song Pad or Guitar

Next, I would bring in the song pad keyboard or whichever instrument plays most of the way through the song and supports the main vocals. This could be an acoustic guitar for a singer/songwriter track or the main electric guitar for a rock band. I balance whatever this choice is on the track well behind the lead vocal in a supporting role, making sure that it is not fighting for space with the vocal; it could always be raised later in the mix process.

Step 3: Main Acoustic Guitar Accompaniment

At this point I consider turning that instrument into a stereo soundscape, if it is not already. The reason for this is to fill out the stereo picture behind the vocals. If the keyboard pad is mono, I put a simple stereo chorus on it (set as subtly as possible) using an auxiliary send. Same for the guitar if it is a mono signal, later in the mixing process I remove that from the acoustic guitar. Currently, when recording a main acoustic guitar for a typical singer/songwriter I generally record three signals into a DAW:

1. A feed from the internal guitar pick up by the direct injection (DI) box.
2. A small diaphragm condenser microphone over the acoustic hole, angling in from the fretboard, typically a Neumann KM84.
3. A large diaphragm condenser microphone on the body of the acoustic guitar, below the guitarist's strumming hand but out of the way of being hit by that hand, typically a Neumann U87 would do this job well or an AKG C414.

Step 4: Lead Vocals

Now I would start processing the lead vocal. Here is my standard set of vocal mix techniques:

1. Insert a vocal compressor starting with a 3:1 ratio and the threshold set so that the gain reduction meter is only active on the louder notes.
2. Set up an auxiliary send to a vocal plate reverb with a decay of about 3 seconds and a high pass filter (HPF) up to 150 Hz.
3. Set up an auxiliary send to a crotchet mono delay effect with around 35% feedback and 100% wet.
4. The equalizer settings are entirely dependent on how the vocals sound but typically if the vocals were recorded flat (and I will not know unless I have recorded them) then I would boost a few decibels (dB) around 10 kHz or 5 kHz and consider a 4 dB cut at 300 to 500 Hz and also a HPF up to 100 Hz, provided this doesn't lose the body of the vocal sound. If the vocal is already sounding too thin then I would try a boost around 150 to 250 Hz, but no lower than that as I would want to save any boost of 100 Hz downwards for kick drum and bass only.

5. My final suggestion on a lead vocal, and I apply this later in the mix, is the Roland Dimension D at its lowest setting: Dimension 1. The Universal Audio plug-in virtual copy of this piece of hardware is a good replacement and again this would be on an auxiliary send, in addition to keeping the original signal in the stereo mix. The effect on the Dimension D is still set to 100% wet and balanced behind the lead vocal to create a stereo spread but strangely it has a wonderful effect of bringing the voice forward, hence this is best used later in the mix, when there is more going on behind the lead vocal.

As may have been gathered by now, I am building a multi-dimensional landscape of sound across a stereo picture. During my mixing process, from the 1980s onwards, I have always used my sonic landscape in relation to a picture landscape. To take that further in the mixing stage I imagine the picture as 3D so that I can analyze the staged layers in a deliberate attempt to separate the instruments for the listener and yet to also help those instruments to mold together as one. This should sound to the listener as though the musicians (live or programed) and singers are in one space – or all on stage – together. I use the word picture deliberately here because that's how I plan a final mix, like a multi-dimensional landscape painting, with the lead vocal at the front and heart of the picture. I usually have this picture in my head before I even start the production and I believe this is a sensible way to plan a commercial pop production. Firstly, you have the vision, which is built around the artist, the direction they and their manager and their label want to go, and then you have the song, either written for or chosen by the artist and the team around them. Then there are various plots, comparisons, influences and directions all chosen by you, the producer(s), in collaboration with the artist, their record label and their manager. That gets thrown into a creative melting pot that you as the producer, engineer and mixer have to deliver, making complete sense and showing that you allowed all of those suggestions to influence the final product. One could compare it to painting by numbers but it is not quite that strict as you are constantly reviewing and changing as you go along. In my experience, the most important person to listen to will be the client, whether that is the artist, manager or more frequently the label. The other thing that made sense in my BoyBand work and this mixing method in the 1990s was that the focus for acts such as East 17, Take That and Boyzone is totally on the boys and their vocals. This is the case whether it is the media (especially radio and television), their manager, the record label or most important of all – the fans. None of these people are focusing on the rhythm or the music, if that is all working well behind the vocals then it is doing its job. This is also the reason we heard so many 1990s BoyBand song introductions featuring vocals, often edited from the chorus that would come in within the first minute of the song. It could be said that what I have described forms my 'Phil Harding Signature Mix Sound' or as Simon Zagorski-Thomas (2014) calls it, 'a schematic mental representation'.

Step 5: Double-Tracked Lead, Harmony and Backing Vocals

The next step is for me to bring in the rest of the vocals behind the lead vocals; any double or triple tracked lead vocals would be 5 dB below the lead vocal. This will give the effect of fattening the lead but without losing the character of the main lead that we are likely to have spent hours editing and tuning after the recording. Typically, this works well in a pop chorus. All of the processing I have described for the main lead vocal would generally go onto the double-track lead vocal except for the crotchet delay, I tend to leave that just for the single lead track, otherwise it can sound confusing if it is on the double as well. Next would be any harmony vocals recorded to the lead vocal, generally a third up or maybe a fifth below. If the harmonies are single tracked then they would remain panned center. If they were double-tracked, then I would pan them half left and right, or even tighter. Processing on these would be similar to the lead vocals but with no delay effects. Finally, to complete the vocal stage of the mixing, we move onto the chorus backing vocal blocks, which would often start with double or quadruple tracked unisons to the lead vocal in the chorus. This is to add strength and depth and a stereo image with these panned fully left and right. From there, all of the other harmonies in the chorus would be panned from the outside fully or, for instance, half left and right for the mid-range harmonies, tight left and right at 10 o'clock and 2 o'clock for the highest harmonies. All of these need to be at least double tracked once to achieve a true stereo. The processing would be applied on the stereo group fader that these vocals are routed to. This saves the computer system DSP by not processing the individual tracks. Typical backing vocal processing would be compression first, set similarly to the lead vocal, equalization, again similar to the lead vocal but less low mid cut and minimal HPF. The vocal reverb would stay the same though it is worth considering a longer reverb time, four seconds or higher to place the backing vocals further back from the lead vocals. I would not put the crotchet delay on the backing vocals except for a special, automated, effect on one or two words. I send the backing vocals to a small amount of a quaver delay overall to give them a different and tighter perspective to the lead vocals. Multiple tests and use of this methodology since the 1990s have proven to me that this is a repeatable formula for all pop and dance mixes. One may wish to vary one's interpretation of this with more delays and processing for extended and club mixes (especially by more use of the crotchet delay on the backing vocals) but for a radio and video mix, the techniques given almost guarantee an industry standard and accepted sound.

Step 6: Pianos and Main Keyboards

Pianos and keyboards add more musicality and support underneath the new vocal stereo sound spectrum that has been created. In terms of time, steps 1 to 5 could take half a day so if I have not already done so, I would take a break before step 6. I would advise taking a break every two hours to rest the ears and equally these days, the eyes, which have been constantly staring at computer screens throughout this process. Some people say that our ears are only good for four hours work per day. I am not sure I agree with this, especially if regular breaks are taken. Often I will be happy to stop my day's work after step 5 and come back completely fresh on another day. I would have already prepared my keyboard stereo group fader (auxiliary input track in Pro Tools in my case) and I would buss all the keyboards at this stage to that group fader. Unlike the lead and backing vocal groups, the keyboard group is unlikely to have any processing, auxiliary sends or inserts, it is just there as an overall level control for typical keyboard overdubs such as synthesizer pads, pianos, organ, bells, and so on. Most virtual keyboard sounds and hardware keyboard sounds will generally deliver a stereo signal and my first job would be panning decisions. One can leave everything panned hard left and right, therefore allowing the original patch programers of those sounds to decide on the stereo image. I prefer to take control of this myself. I will even go as far on a programed acoustic piano part to split the stereo audio track into individual left and right mono tracks. I do this so that I can process each side of the piano individually. This raises an important point that applies to drums as well as pianos. One needs to make a decision on the stereo image of some instruments: Are we panning as though the listener is the performer/player or are we panning as though the listener is the audience, like in a live show? My preference is the performer's perspective, therefore you imagine you are the piano player in this case and your left hand, or low end piano part, is panned hard left and your right hand or mid to high end piano part is panned hard right. This puts the listener in the piano seat, a wonderful perspective in my view and particularly effective for the listener on headphones. Therefore, when we get to them, I recommend the same panning perspective on the drums. Put the listener on the drum stool. For my typical equalization for a stereo acoustic piano, this is the same whether it is a live piano or a programed stereo sample, I generally engage an LPF down to around 8 or 7 kHz on the left hand side and then a small boost around 3 kHz to bring out the rasp of the low piano notes. Then, coming down the frequency spectrum I consider a 2 to 6 dB cut in the low mid frequency (300 to 900 Hz) to eliminate any unwanted 'boxy' sound, as I call it. Finally, I may want to boost some low end to bring out the depth of the piano, I would boost this from 150 to 200 Hz as I am saving the frequency ranges below that for the bass (100 Hz) and the kick (50 Hz). I know all this sounds very strict and specific but it works as an efficient start point. For the right hand side of the piano, I do almost the opposite to the processing of the left hand. I start with an HPF up to around 150 Hz, perhaps higher, I engage a small low mid-frequency cut between 300 to 900 Hz again but it could be left flat if one prefers. Then I experiment with boosts around 5 kHz and 10 kHz, with large bandwidths to brighten the piano. I find that when you bypass these left and right equalizers and then put them back in for comparison, that width and separation of the stereo perspective of the left and right hand piano parts will be enhanced. The other aspect of having split the stereo piano track into mono left and right is that after all of the equalization I have described, you can add a little of the vocal reverb plate to the right hand side of the piano only, this will help to balance the piano into the track and bring it closer to the vocals. The final thing to consider on the piano is compression, which ideally should be first in the insert chain, so as not to be affected by your equalization choices. Again, I keep the compression simple starting with a 3:1 ratio and adjusting the threshold to activate the gain-reduction on the louder parts.

Step 7: Other Keyboards and Orchestration

Other keyboard parts have to be treated on their own merits and ideally one should find a space for them that either fits behind the main piano or pad, or they jump out as a feature of their own. If there are programed string and orchestra parts I would deal with them next. Typically for strings I like to have a longer plate or hall reverb setting, at around 4 to 5 seconds in length and ready to go on an auxiliary send. I would have an HPF on the strings up to 150 Hz and definitely no LPF, I prefer a completely flat top end. For brass, programed or live, I would send them to a small plate or a room setting at around 1.5 seconds in length. Other rhythmic keyboard or sequence parts I leave until I add the drums but certainly I would check them at this stage for any need to add compression and equalization, also possibly rhythmic quaver delay to bed them into the track. It is not unusual to have strings and orchestra on pop productions, whether they are real or programed samples, the mix processing on both are similar. For the orchestra hall reverb, where I have already explained the settings, I would only send the violins (first and second) to this, plus the harp (often arranged to work with the strings) but a lesser amount on the violas. I would generally keep celli and double basses dry. If there were woodwind as well, I would send them to the vocal reverb plate so that they are a little lighter than the violins. My ideal string recording setup is described in Howard Massey's book, *The Great British Recording Studios* (Massey, 2015, p.178)[5]. My equalizer recommendations on a live orchestra would be:

1. An HPF on the violins to cut out celli and double bass spillage up to 300 Hz, then a small boost between 8 to 12 kHz.
2. For the violas I would use a HPF up to 100 Hz and a small top end boost at 5 kHz.

3. For celli I would generally leave them flat other than a small boost at 3 kHz if they need it to cut through the balance.
4. The double basses I would keep flat.
5. For the harp I would consider a HPF up to 200 Hz, a small cut at 300 to 900 Hz and small boost at 8 to10 kHz.
6. For the woodwind, a small boost around 5 to10 kHz and a HPF up to 100 Hz.

All of this should help the orchestra to merge together and to blend into the track. Notice there has been very little low middle cutting on the orchestra in my loathed 300 to 900 Hz frequencies.

Step 8: Guitars: Acoustic and Electric

I would generally deal with acoustic guitars before electric guitars. If the production were centered on an acoustic singer/songwriter then the artist's main guitar part would have been my first support instrument of choice whilst processing the lead vocal. I described in step 3 how I would record an acoustic guitar for a singer/songwriter orientated production and I would now record any acoustic guitar part the same way for safety but in a double or triple tracked guitar backing I would only use the microphone that was on the guitar sound-hole (Neumann KM84). The equalization choices for the acoustic guitar on the mix would typically be HPF up to 100 Hz; a 2 to 6 dB cut at 300 to 900 Hz and if it sounds too dull then apply boosts at 5 kHz and 10 kHz. All of this is for the multitracked one microphone acoustic. If you have just one main acoustic and you are using the multi microphone technique and direct injection that I described in step 3, then I would pan the 'body' condenser signal hard or half left and leave it virtually flat other than a 300 to 900 Hz cut, I would even consider a 150–200 Hz boost or 2 to 4 dB to give more depth. I would then pan the sound-hole microphone hard or half right and duplicate the equalization described for the multitrack acoustic but possibly with the HPF up to 200 Hz. The D.I. signal would feed in behind the stereo microphones in the middle, probably kept flat but at this stage one should check phasing of the three signals combined and finally consider a tiny bit of the vocal reverb plate on the right hand signal to help blend the acoustic guitar into the track. Certainly I would avoid room, ambience or hall reverb.

Electric guitars are so technically varied these days I would again record three signals if I were involved at the recording stage:

1. A dynamic microphone such as the Shure57.
2. A large diaphragm condenser such as the Neumann U87 or AKG414 or even a ribbon microphone on the guitar speaker cabinet.
3. A D.I. signal from the guitar for any future re-amping plug-ins to be added at the mix stage.

In the hope that the guitar sounds are well sourced and well played, I do little or nothing at the mix stage. I generally do not touch the low frequencies or lower mid frequencies, deliberately leaving them in because I am cutting them so much elsewhere. If the guitar sounds at all dull, I only boost around 4.5 kHz as I am trying to leave 5 kHz and above to the vocals, piano, acoustic guitars and cymbals on the drums. I would only consider any reverb or delays if the guitarist has not used any pedals or guitar amp effects. Thankfully, the technology of guitar effect pedals and the effects on guitar amps is good enough now, I believe, for engineers to trust their quality and low noise ratios. I generally trust the guitarist to deliver the sound that feels right to them on their amplifier to suit the track. If the guitars have been recorded flat and dry I would add some short plate or ambience reverb between 1 to 2 seconds in length and some quaver delay with around 30% feedback for rhythm guitars. Usually I would apply some longer reverb and crotchet delay for solo and lead guitar parts.

Step 9: Bass

Much like the electric guitars, I do very little to the bass. I would hope to have D.I. and amplifier signals and I would use a single large diaphragm condenser microphone on the bass amplifier at the recording stage (Neumann U87). I compress the signals at the first stage of the mix insert chain, being careful to only use 2:1 ratio as anything above that can destroy the low frequencies. I generally boost both signal paths at 100 Hz (I have been saving this frequency in this 'Top Down' mix method, exclusively for the bass). Finally, this is where I would try the Roland Dimension D using the same settings and auxiliary send set up as I did on the lead vocal. This will add a stereo perspective to the bass and some more warmth. I also add a small boost around 1 to 2 kHz (the only time I use these frequencies), after the drums are in, if the bass is not cutting through the mix or sounds as if it needs more edge. Regularly now I will also experiment with the Sansamp on the D.I. signal, this is a useful amplifier simulator plug-in that comes free with all versions of Pro Tools.

Step 10: Drums: Live or Programed

Finally we get to the drums, which for the traditional rock, pop and dance genre mix is usually the first step. What I have found strange and yet enlightening in my forty-year career is the similarity in the methods and sometimes even sounds for drums, certainly the processing of them, which can work for all three genres: Rock, pop and dance. Generally, metal rock drums will need a different specialist approach but it is astounding that similar compressions, gating, equalization and ambiences will all work similarly to give drums the same amounts of power required. At this point I have only roughly balanced the various elements from the vocals down to the bass but the important thing is that I will have looked at everything individually. I will regularly need to mute the vocals to achieve some of the things I have described here but I will put the vocals back in as I move through each step. Now for the drums, I need to focus mainly on bass, drums and guitars in the balance. It is difficult to start gating, compression and equalization accurately on drums when the vocals are still in the mix and prominent. As with a 'bottom up' mix I would start with the kick but I usually bring the whole kit in roughly and quickly under the current balance to remind myself of what the drums are doing. In my chain of kick inserts would be gate first, then compression, then equalization. I prefer to use inserts in that order, as I do not want the equalization to affect my gating and compression settings. I would set up a side chain if I felt that was required. Copying the kick track before one starts the processing is a good idea because the gating and equalization can end up being drastic (especially on live drums) and you may wish to balance in the unprocessed original track. The same can be done for the snare track. Typical equalization on the kick would be a 2 to 6 dB cut at 300 to 900 Hz, a boost of 2 to 6 dB at 50 Hz. If more 'slap' or pedal-beater is required then boost 2 to 6 dB at 3 kHz but no higher than that as we are saving 4 kHz and upwards for the guitars, keyboards, strings, vocals and cymbals. Typical snare equalization would be a 2 to 6 dB cut at 300 to 900 Hz, boosts of 2 to 6 dB at 4.5 kHz and 2 to 6 dB at 7 or 8 kHz. The kick and snare gating would aim to isolate them from the rest of the drum kit and typically one would hope to eliminate snare spill on the kick signal and hi-hat spill onto the snare signal, always being careful to allow as much release and hold time to allow the drum decay to breathe before the gate closes. Compression on the kick and snare should be minimal as they will be likely to drive compression on the overall mix at the end of the session. An individual compressor on the kick should start at a 2:1 ratio for the same reason as mentioned on the bass; higher compression ratios can have the effect of losing low frequencies. A major part of the snare sound will be the choice of room ambience effect, even something like the D-Verb in Pro Tools, set to medium room and below one second in length, can give one a good room ambience for drums and instantly the snare will be more powerful and sound as though it is part of the track. I would also send the tom-toms to the same room ambience effect, then either gate the tom-toms to eradicate spill from the rest of the drum kit or more efficiently one could edit them to only remain where the tom-toms play throughout the track. Any equalization on tom-toms would be minimal, a moderate boost around 4.5 kHz is all I would do if they are lacking attack or sounding too dull. Ideally, overhead cymbals and hi-hats should all have similar equalization to emphasize the cymbal sound and to eradicate kick, snare and tom-tom spillage. My equalization for these would be a HPF up to 200 to 300 Hz, a cut of 4 to 6 dB at 300 Hz to 900 Hz, a small boost at 5 kHz and 10 kHz.

That concludes the drum processing. An important decision is panning perspective, as was the case with the piano. We need to decide on panning to the 'player perspective': Hi-hat left, tom-toms left to right, overheads left to right. Or the 'audience perspective': Hi-hat right, tom-toms right to left and overheads right to left. As aforementioned, I much prefer the player version on drums and piano. If one has stereo room microphones recorded for the drum session make sure that their panning matches the close microphones and consider a similar equalization to the overhead cymbals. I often experiment on the room microphones with some heavier compression at a 6:1 ratio or more and an overdriven threshold setting. One will need to raise the compression output or make-up gain but this can sound powerful. My personal microphone choices when recording drums are:

1. Kick: Electrovoice RE20.
2. Snare: Shure SM57 (top and bottom).
3. Hi-hat: AKG C451.
4. Tom-toms: Beyerdynamic101s.
5. Overheads: AKG C451s.
6. Room microphones: Neumann U87s.
7. Ride cymbal AKG C451 on the bell of the ride cymbal and route that to its own track to enable control of this during guitar solos and so on.

Step 11: Final Stereo Balancing and Tweaks

Everything has now been individually processed and each step has dealt with a group of overdubs or a single instrument at a time. Hopefully after the drum step, the main lead vocals should still be prominent in the balance. Now is the time to achieve the overall mix balance and if I have been listening on large studio monitors or loudly on nearfield monitors – one often has to when processing individual sounds – I would consider taking a break. I believe it is our duty as creative technicians to make sure that the overall mix is playable without ear damage at high volume. This is because many listeners, hopefully excited by the music we create, want to listen loud. But at this stage of the mix one should come down to a quiet or medium volume on your nearfield speakers to achieve the final stereo balance. In an ideal world I would approach step 11 on a fresh day, or at least after an extended break. I find that these final tweaks can take up to half a day, maybe checking on headphones, flipping between speakers and volumes and possibly running a test mix wav file or CD to listen to in different environments such as your home hi-fi system or car stereo. Also, at this final step, I would send the lead vocals and bass to the Roland Dimension D, as mentioned earlier, to help the lead vocal stand out from the track; the Dimension D has an effect of bringing the vocal forward. Whilst we do not want the bass to be as loud as the lead vocal, the Dimension D effect brings it forward and gives a stereo perspective that is satisfying.

Step 12: Master Bus Compression

Overall limiting or compression, subtly, on the stereo mix master or buss is the last thing to look at. A low compression ratio of only 2:1 or 3:1 is a good start point and the idea of limiting or compression here is only to subtly control any peaks in the overall mix (not to boost the level, that is a different process). We should always bounce or export an unmastered 24bit wav file for future mastering. For a listening copy to go to the clients and artists, I would now insert a digital compressor/maximizer such as 'Maxim' in Pro Tools. This can raise the overall mix level to existing commercial levels[6], which means that should the clients or artists compare the mix to another commercial release, the mix will be close to that kind of volume and peak level. See the Maxim screenshot in Figure 6.1 for one of my typical settings.

As a first mix, I would generally expect clients and artists to come back with comments and suggestions that will require a revisit to the saved multitrack file and session. For this is now common industry practice, it is the reason I have generally dispensed with using console automation on commercial pop tracks that are mixed entirely

Figure 6.1 Typical Maxim settings, inserted on the stereo Master Fader in Pro Tools.

in the DAW. If I have achieved the whole mix 'in the box' with no outboard hardware, recalling the mix is fast and easy. Hopefully the requested adjustments will also be fast and easy, whether you agree with them or not. If, for instance, I have automated the lead vocal on the initial mix, it becomes awkward to make level adjustments on other instruments because everything would still need to work with the potentially rigidly automated lead vocal levels and rides. This is difficult to explain but will become obvious with continued practice. The way around this is to go through the lead vocal track, zooming in on the wave display, then go through the song and make the required adjustments, up or down, by using the audio 'gain rendering'. I know this sounds a little strange and even unprofessional but continued practice for me became second nature and by applying mix adjustments of 4 to 6 dB or higher (common in pop), it became easier and I found myself no longer battling automation left over from the first mix. The only things I tend to automate now are extra vocal delay-send boosts (to emphasize one or two words) or quick equalization changes to help any vocal 'pops' that have slipped through on the recording.

Final Notes for the 12-Step System

The idea with this method of mixing is that one is concentrated on the song from the minute of starting the mix and although I have described the process from the vocals down, it would also work for instrumentals and other non-vocal orientated tracks where you would start from your lead instrument or 'theme'. Fauconnier and Turner (2002) talk about 'conceptual blending' as an unconscious human activity and that is exactly what I seem to do in my mixing approach. Short ambience on the drums blended with medium to long plate reverb on the vocals, then those vocal reverbs blended with my crotchet mono delay as highlighted in step 4. These are good examples that describe the types of technology we engage whilst mixing a record using today's technology. Another common industry practice in 2016 that I have not mentioned is the Dropbox file sharing system. This is useful when either engaging others to perform a mix for you or also if you have been hired for a mix. Exchanging multitrack files through Dropbox has become common practice and it is recommended to sign up for an account to enable participation in this.

Technical Guidance: Export and Import Procedure to a Mix Template File

My current (2019) pop production technology process is to start compositions and productions in Logic Pro. At the point of a completed production, I then export all of the individual tracks, both MIDI and audio, that are still open in the final session file, as wav audio files 24bit/44.1 kHz (or 48 kHz) or higher as the project permits. Before doing this, for efficiency, I would go through all of the tracks left to right (L–R) on the mix or edit window screen and add a number from 01 onwards to all of the track names. This means that when you or someone else imports them, all of the tracks will line up L–R in numerical order on any typical music software mix window. This is a fast and efficient process in Logic Pro. Highlight all of the open tracks from bar 1 beat 1 (this is standard industry practice) to the end of the song. From the software pull-down menu select 'File>Export>All Tracks as Audio Files' and be sure to bypass Effect Plug-ins and do not include Volume/Pan Automation. See the screenshot in Figure 6.2. This will now be exported as raw audio data, ready to be imported into Pro Tools or any music software of your choice. It would also be useful to export one MIDI file from Logic Pro (a pad or piano I would suggest) to retain the song arrangement markers and tempo; this saves a lot of time in Pro Tools.

   I would suggest that before importing this into a Pro Tools session some time is spent creating a blank (no audio or MIDI data) Pro Tools mix template file (see the screenshots in Figures 6.3 and 6.4). This is a useful time saver. It is a tiny file that each time you are ready to start a new mix, you create a new folder, copy your mix template Pro Tools file into it, launch Pro Tools from that file, rename the song in a 'save as' from the file menu option and start by importing the MIDI track from Logic Pro. Position and highlight the track to the left of the auxiliary groups (refer to the template screenshot in Figure 6.4) and then import all of the audio tracks, also at bar 1 beat 1. The final task before starting the 12-step mixing program is to position all of the stereo auxiliary bus faders to the right of the cluster of audio tracks you want them to control. I route those audio tracks, through busses, to the corresponding input busses of the

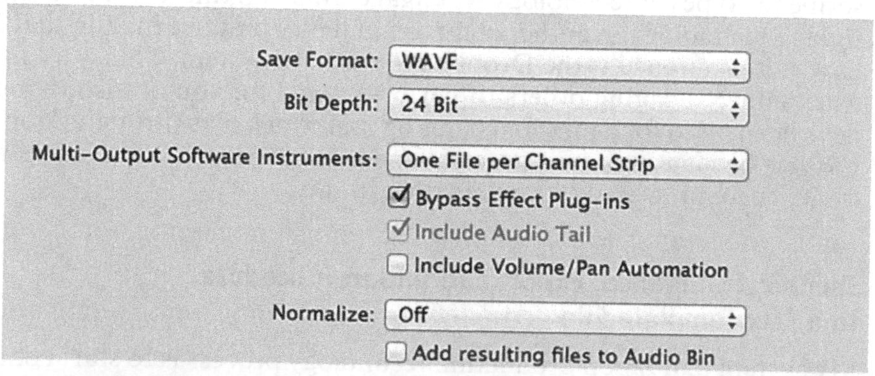

Figure 6.2 Logic Pro: Export 'All Tracks as Audio Files' menu choice recommendation.

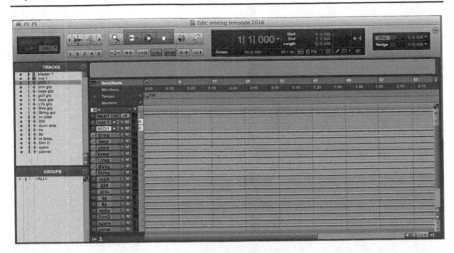

Figure 6.3 Pro Tools Mix Template edit window.

Figure 6.4 Pro Tools Mix Template edit window.

Figure 6.5 The Phil Harding E.Q. chart.
Illustrated by Vince Canning, 2018. Copyright Phil Harding 2018.

stereo auxiliary bus faders, for instance, all drums and percussion can go to buss 11–12 (we have saved busses 1–10 for auxiliary effects sends). See the Pro Tools mix template screenshots in Figures 6.3 and 6.4 for how this final process should look.

Now refer back to Step 1.

As a reference guide to the EQ examples throughout the 12-Step program, I have developed the Commercial Pop Music EQ guide chart (see Figure 6.5).

NOTES

1. Commentary on the Jamie Shaw project and orchestral recording methodology are in the Appendix.
2. Descriptions of the three commercial Strongroom Studios throughout the 1990s can be found in the Appendix.
3. In my discussions throughout these steps I will refer to the 300 to 900 Hz frequencies as disliked or loathed by myself. This is due to the fact that I have never found an occasion when it is useful to boost this frequency range. I find it to be 'muddy and boxy' when boosted and will generally recommend either cutting these frequencies or leaving them flat.
4. There is a technical guidance section at the end of this chapter that describes my audio and MIDI file export procedure from Logic Pro to Pro Tools. You may want to read that before embarking on step 1.
5. Also in the technical Appendix.
6. New loudness normalization standards in streaming services of approximately −14–16 LUFS may become the norm.

Pop Record Examples

7.1 INTRODUCTION

The following pop record commentaries cover a range of my music industry work coming out of the 1980s pop production and songwriting period in the published book, *PWL From The Factory Floor* (Harding, 2010). I concentrate on my 1990s single and album co-productions (with Ian Curnow) of BoyBands such as East 17 and Boyzone. Concluding this chapter is commentary on the 1999 Channel 4 television popumentary 'Boyz Unlimited', which takes a cynical, fictional view of the 1990s BoyBand phenomenon.

7.2 'HOUSE OF LOVE' (SINGLE) BY EAST 17

#11 UK Charts, August 1992

East 17 Members:

Tony Mortimer: Principle songwriter and rapper.

Brian Harvey: Lead vocalist.

Terry Caldwell: Backing vocalist and dancer.

John Hendy: Backing vocalist and dancer.

Songwriters: Tony Mortimer and Robin Goodfellow.

Producer: Robin Goodfellow.

Remix and Additional Production: Phil Harding and Ian Curnow.

Manager: Tom Watkins.

This commentary of the first East 17 single, 'House of Love', is an example of early 1990s BoyBand production. The record was a remix and additional production credit for Ian and I but the result quickly became a benchmark for London Records that P&E could produce a cutting-edge sound of the moment that would progress with our teamwork, led by Tom Watkins, to rival any early 1990s BoyBands. This work and the sociology surrounding it allow the opportunity to look at some of Pierre Bourdieu's theories in practice. By 1992, with Tom Watkins, East 17's manager, we had a multimedia entrepreneur whose experience and knowledge of the pop industry and pop culture was second to none. Tom had steered the Pet Shop Boys to international success throughout the 1980s and then groomed and manufactured Bros to British and European success by the late 1980s. His *cultural capital*, as Bourdieu (1984) calls it, was enormous and having persuaded London Records to sign his latest act, East 17, Tom Watkins had created a commitment to London Records to deliver on the *economic capital* that they were investing in him. London Records trusted that he would succeed because of his *symbolic capital*, his status as a successful entrepreneur, and was deemed worthy of their investment. Throughout the 1990s, Tom Watkins would always want to be totally in control of every aspect of his creative projects, from the image and design around the artist through to the final mix and the promotion of the product.

I interviewed Tom Watkins in May 2014 and was surprised to hear him still praising the work that Ian Curnow and I did on the first East 17 hit single, 'House of Love'. Ian and I would go on to produce and mix bigger hits for East 17, under Tom's guidance[1] but it was still this first collaborative work that Ian and I did with him that has stayed prominently in his mind. Even when I suggested during the interview that he reflect on the further success of tracks like the #1 single 'Stay Another Day' Tom was adamant that this track contained our best work. In his own words:

> When we worked on the [first] East 17 album I remember Tony Mortimer coming down and you had just done 'House of Love' which I still think to this day is a fucking amazing record. I remember sitting in the Strongroom and listening to it in the big room and thinking 'fuck me' you know, this is [brilliant?] As did Tracy Bennett [London Records A&R], and I remember Tony Mortimer turned round and said 'thanks for fucking up my record'. Do you remember that? I looked at him and said 'you are so wrong'. I know in his mind, he thought he was the new messiah but I knew and you knew that it was you and Ian sitting in that fucking control room that created that record.
>
> (Watkins, 2014, personal interview)

Tom's comment with regard to 'House of Love' and how Ian and I 'created' that record are further underlined by songwriter and producer, Mike Stock, in Chapter 3, where Stock describes pop producers as being 'the band'. Producer, Robin Goodfellow, originally recorded 'House of Love' at around 114bpm in a hip-hop style, earlier in 1992. When Tom Watkins came to P&E for the remix he wanted us to increase the speed to be closer to 120bpm and re-produce the track, turning it into a 'Frankie Goes to Hollywood' parody. That required a large transformation of the production by Ian and I and here is my commentary of that process.

In early 1992, no music sequencer program manufacturer had so far achieved a solution that many music producers craved – to incorporate MIDI programming (Musical Instrument Digital Interface) with audio recording in one package. P&E were lucky that for many years Ian Curnow had been a beta-tester for Steinberg's Cubase software and at this time in the summer of 1992, only MIDI programming on the Atari ST computer (designed for computer gaming but it usefully had two MIDI ports built-in) was available for Cubase and Logic[2]. Steinberg were developing what was required to incorporate audio into their MIDI program so that recording studios, songwriters, producers and remixers, such as P&E, could move away from the analogue or expensive multitrack digital tape format and onto computers for digital audio recording. It became clear that the Atari ST computer was never going to be powerful enough to cope with the required amount of digital sound processing (DSP). Therefore, a different or new computer base was required to move forwards with this technology. In the early 1990s, American company, Apple Macintosh, had been collaborating with another music recording and sequencing package by Digidesign called 'SoundDesigner'. This was only ever developed to record stereo audio and it would be a long time before Digidesign incorporated MIDI into their audio programme. London-based American computer programer, Mark Badger of Steinberg, had been transferred to Silicon Valley in California to collaborate with other music design companies, such as Digidesign and Apple. It became clear to Mark that the way forwards was to move the Steinberg Cubase MIDI sequencer onto the Apple Macintosh computer platform and collaborate with Digidesign for their audio recording expertise, combining these two into the Cubase program so that music programers, such as Ian Curnow, could seamlessly record audio into the same type of track recording as their musical MIDI parts. This was a truly exciting innovation for producers, music programers and remixers. Ian and I were lucky at the time of embarking on the 'House of Love' remix that Mark Badger and Steinberg were at the beta-testing stage for this new version of Cubase that would initially incorporate four tracks of audio recording with traditional MIDI recording. Mark Badger agreed to come into The Strongroom Studios and assist Ian and me with the task of taking 'House of Love' from its 114bpm hip-hop style to the required 120bpm. The start point of this process was to find a vari-speed on the analogue multitrack tape machine that would lift the track a semitone up so that Ian could program new keyboards to the existing vocals and a few other key audio tracks that we wished to keep for the final mix.

Mark Badger assisted Ian with this task whilst I was in studio 1 working on the vocal balances and processing. The vocals were sampled into the beta version of Cubase on the Apple system, allowing Ian to begin programming the new keyboards and drums at this higher tempo and key to develop the track to sound more like Frankie Goes To Hollywood (FGTH).

   This took us over two days and most of the night working through because once Ian had programmed the new music sequences he would route the overdubs down to studio 1 for me to record back on to the 48-track analogue tape machines. Ian and I realized that this was going to work creatively and fulfill the request from Tom Watkins to have a fours-based kick pop track in the FGTH style. When Tom Watkins and his partner, Richard Stannard, came to the studio on Monday morning to hear the newly worked version of 'House of Love', they were stunned, bewildered and delighted that Ian and I were able, in a fairly short space of time, to turn the track into almost exactly what Tom wanted to hear. Tom and Richard were not particularly interested in the technical detail of how we got the result but it was clear to me that the creative technical tasks that Ian and I had set ourselves over the weekend were worth the time the hard work and the lack of sleep. Technically, this also gave Mark Badger the evidence to present to Steinberg that in a practical studio situation this was going to be a major breakthrough for the Cubase program. Steinberg would become, for a number of years, the market-leaders in this newly integrated technology, which today we take for granted in our Mac and PC systems. Laptops and tablets are also now capable of combining MIDI and audio recording. Ian Curnow stated:

> This was the beginning of the change in the way we all made records. It was developed from SoundDesigner and had a two track [stereo] digital sound-editing program.
>
> (Harding, 2010, p.557: Interview by
> John Palmer, July 2010)

In my interviews in 2014 with Tom Watkins and with East 17 founder and songwriter, Tony Mortimer, it is clear that there were differing opinions between Tom and the band. Tom was delighted with the transformation of 'House of Love' but it was some time after that that we met East 17 and we were told that actually, they were not happy with our mix. Interestingly though, Tony Mortimer does now admit that our mix and transformation of 'House of Love' turned the key and opened the door to the possibility of East 17 becoming a successful BoyBand, which was Tony's original desire. Ian and I completely restructured the song arrangement from their original recording, we retained the three verses and the opening synth riff plus the two vocal chant sections, the track's length was 4'42" and this was largely to please the managers, Tom Watkins and Richard Stannard. They wanted to hear the track with an exciting introduction build and the chorus at the beginning, as well as the original chant. How the unusual arrangement was constructed is detailed in the following paragraphs.

**East 17: 'House of Love' Pedigree Mix (album version)
approximately 120bpm – 4'42"**

Intro x 8 bars: After some dog barking samples there is an atmospheric, film soundtrack type of dramatic build, featuring synthesizer pads and voice sample sounds that climax with mock electric guitars[3] from Ian Curnow that lead into an introduction chorus with a kick drum into echo marking beat 1 of the bar.

Intro Chorus Chant 1 (breakdown) x 8 bars: This chant only appears once in the whole arrangement and is a slower pace than the main choruses and the second chant that appears after the third chorus.

Intro instrumental riff x 8 bars: Apart from all the vocals, this was the only other overdub that Ian and I retained from the originally recorded version. Even then it only lasts for 4 bars as we felt the need for another 4 bars of dramatic 'FGTH-style' build to take us into a full intro chorus. A USA police car siren blasts throughout this to create a sense of tension. Everything else is played and programmed by Ian Curnow. Ian recalls:

> I remember it was a nightmare getting our drum loops, stored in the Akai [1000] sampler at that time, in time with the track because at that time to get loops in time you had to tune them inside the Akai and I had a reference list of tunings against bpm and for some reason it was difficult on this track because of the varispeed process we had gone through on the tape transfer. A lot of trial and error took place to be sure the loops stayed in time on a bar to bar basis. Some of the loops we used in the verses came from Richard Stannard [Tom Watkins partner at the time] so they weren't part of our regular Akai library – which were all set to 120bpm default by the way.
>
> (Curnow, 2014, personal interview)

Intro Chorus x 4 bars: Half of a full chorus that allows us to introduce the main hook that will come a further three times in the song.

Verse 1 x 8 bars: Finally the track settles into a rap verse from Tony Mortimer that just about survives being sped up around 5bpm. The atmosphere and loops behind it come from Richard Stannard's suggestions, as stated by Ian Curnow.

Bridge 1 x 8 bars: Brian Harvey's vocals introduce an aggressive bridge that builds into the chorus. There are some powerful orchestra sample stabs and dramatic choir lines throughout the 8 bar bridge. Typically, as a production technique, Ian and I drop out the rhythm for the last bar of the bridge to give the downbeat of the chorus maximum impact, aided by Ian's 'Eddie' guitars at the end of the bar that run up into the first chorus line.

Chorus 1 x 8 bars: The whole band joins in vocally with the chant-like chorus and Ian Curnow lifts the choir and orchestra samples to an inversion higher to maintain the drama in the chorus, which has the tried and tested formula of 2 lines going up, answered by two lines (mainly Tony) going down to contrast the main vocal hook.

Verse 2 x 8 bars: This verse is a repeat of verse 1 but with different lyrics and some extra explosion samples from P&E to match the vocal parts at the end of the verse.

Bridge 2 x 8 bars: A repeat of bridge 1 with the same lyrics.

Chorus 2 x 8 bars: Direct repeat of chorus 1.

Middle Break x 4 bars: The same USA police car siren that was heard across the introduction is featured here together with a dramatic choir sample line. This provides a contrast and relief for 4 bars with the bass also muted for maximum dynamic affect.

Verse 3 x 8 bars: Repeat of verse 1 but different lyrics.

Bridge 3 x 8 bars: Repeat of bridge 1 with the same lyrics.

Chorus 3 x 16 bars: Direct repeat of chorus 1 but repeated twice. It is likely that I edited this back down to 8 bars for the radio version.

Chorus Chant 2 (Breakdown) x 8 bars: This chant is faster than the introduction chant, it still reflects the chorus hook but has even more pace and energy that both chant 1 and the main chorus vocal hook.

Chorus 2 x 8 bars + Ending: Direct repeat of chorus 1 but with an explosion in the last bar replacing the word 'Love' and the fall out from the explosion provides a dramatic stopping point for the track. Tom Watkins insisted that the track end like this and not on a fade.

Having accepted many remixes and a few productions since Ian and I departed from PWL, in August 1992 came a turning point that would mark the P&E journey for the next three years. Ian and I agreed that Tom Watkins would be our manager and we began production on two tracks for the first East 17 album, 'Walthamstow'. These were 'Gotta Do Something' and 'Gold', plus Ian and I remixed the third single 'Deep'[4]. During our holiday break at the end of August 1992, East 17s 'House of Love' entered the top 20 UK charts and eventually peaked at #11. The demand and pressure for a follow-up hit single was tremendous from Tom Watkins and Tracy Bennett (A&R at London Records). I have called that 'follow-up-itis' in my book, *PWL From The Factory Floor* (Harding, 2010). 'Gold' was East 17's second single and it performed disappointingly in the UK charts but a better single 'Deep' had already been recorded during the initial sessions with producer, Robin Goodfellow. Tom Watkins knew that this was going to be a big single, even though it was a different style to 'House of Love'. His confidence was another example of his power in the *domain* and his *cultural capital* in knowing what would succeed in the UK Charts in 1992.

Conclusion

There was an unusual business conclusion to the 'House of Love' project. Ian and I decided to ask for a production royalty *after* the record had been a UK hit. This was unheard of as a remix and production team because the business deal is always negotiated in advance of the project. Ian and I did this because:

1. We had worked so hard in transforming the record and it had become a hit, Ian and I genuinely felt that we deserved to benefit from the sales.
2. We were testing Tom's management abilities, in terms of representing P&E and fighting our corner as producers and remixers.
3. Ian and I were not concerned as to whether we carried on working with East 17 at that time or not, such was the self-belief in our production and remix abilities.

Tom phoned Ian and I not long after our request and said that he had managed to negotiate 0.5% royalty for us with London Records and the contract to confirm that was on its way. Ian and I were genuinely surprised, though happy that our contribution to the success of 'House of Love' was being acknowledged in this way. It put Tom in a good position with regard to our management agreement from that point onwards. Unusually for the time, there was no contract ever signed between Massive Management and P&E Music, just confirmations by fax and a gentlemen's handshake.

Tom's *cultural capital* pulled him through, his *symbolic capital* was retained once the record charted and his *economic capital* promises to the record company, East 17 and to P&E were fulfilled. This was the beginning of a three-to-four-year journey for all of us as a creative team.

7.3 'STAY ANOTHER DAY' (SINGLE) BY EAST 17

#1 UK Charts, Christmas 1994, remained at #1 for 5 weeks into 1995.
East 17 Members:
Tony Mortimer: Principle songwriter and rapper.
Brian Harvey: Lead vocalist.
Terry Caldwell: Backing vocalist and dancer.
John Hendy: Backing vocalist and dancer.
Songwriters: Tony Mortimer, Dominic Hawken and Rob Kean.
Producers: Phil Harding, Ian Curnow (P&E Music) and Rob Kean.
Manager: Tom Watkins.

The commentary of the P&E production 'Stay Another Day' by East 17 describes a creative and cultural process between manager, Tom

Watkins, producers Phil Harding, Ian Curnow and Rob Kean, the artists, East 17, and their record company, London Records. It had been two years since East 17's first UK hit 'House of Love' and a string of UK top 10 and top 20 singles followed, as well as the successful debut album 'Walthamstow'. In his 2007 book, *Group Genius*, Keith Sawyer calls this *group flow*. He states that '*group flow* is a peak experience, a group performing at its top level of ability'. By the summer of 1994, during the recording of this single, every link in the creative and commerce chain was in place and working, from the songwriting genius of Tony Mortimer all the way through to the promotion team at London Records and the independent promoters Tom Watkins had persuaded London Records to hire. There comes a defining moment in every successful music producer's career and sometimes you may feel it at the time of the *flow* of experience and sometimes it may be during a process of reflection. Certainly self-reflection allows a sense of *Being and Time* (Heidegger, 1962, 1927), which Heidegger describes as:

> Such entities are not thereby objects for knowing the world theoretically, they are simply what gets used, what gets produced and so forth.
>
> (Heidegger, 1962, 1927, p.95)

East 17's 'Stay Another Day' entered the UK singles chart at #7 on 10 December 1994 and then rose to #1 the following week, achieving five weeks at #1 in the UK singles chart, including being #1 for Christmas 1994. This record was the fourth biggest selling BoyBand single of the 1990s and the only UK #1 that East 17 achieved. By being the 1994 Christmas #1 it has set the group firmly in the UK pop hall of fame and Tony Mortimer won the prestigious Ivor Novello Award 'Songwriter of The Year 1994' for the song. Ian Curnow and I did not study any theory of 'how to produce a #1 Christmas record', we simply entered into this production at the time with a sense of knowing we could make this track as orchestrated and over-produced as we wished because we had permission from the songwriter, artists and their manager to do so. The introduction of the peeling tubular bells at the end of the track was simply a matter of allowing the process to *flow* and discovering something familiar to people. Mihaly Csikszentmihalyi (1997) describes the *flow* of creativity as 'a subjective phenomenon' and in a way that aligns with this project's creative *flow*:

> All it takes to be creative, then, is an inner assurance that what I think or do is new and valuable. There is nothing wrong with defining creativity this way, as long as we realize that this is not at all what the term was originally supposed to mean – namely, to bring into existence something genuinely new that is valued enough to be added to culture.
>
> (Csikszentmihalyi, 1997, p.25)

Ian and I possessed an inner assurance and confidence that we could create something new and valuable with this production and it was clear as soon as we heard Tony Mortimer's song demo that 'Stay Another Day' had the potential to become a very special record. The song demo had Phil Collins-style drums throughout, which Ian and I thought unsuitable for a gentle ballad about wishing that someone would stay in this world for one more day (the song is allegedly about songwriter, Tony Mortimer, visiting a dying family relative in hospital and wanting them to 'stay another day') but Tony was insistent that these drums should remain.

As the team musical arranger, Ian Curnow's first job was to unravel the chords and arrangement to look for musical improvements that would allow for a more sophisticated production. Once Ian and I had analyzed the song arrangement, programing began on the initial production parts to sound as commercial for radio as possible. For P&E, this was part of our pop production methodology, to squeeze every commercial hook out of the song and maximize the catchy phrases and sections. This was something that Ian and I had learned to harness whilst working for Pete Waterman and the SAW production team in the 1980s. Ian and I realized there was potential for a radio-friendly vocal chorus to enhance the song commercially. Beyond that we were not sure how far Tony Mortimer wanted Ian to take the orchestral arrangement and how much to expand that at the end of the record. Tony's answer was 'go for it guys – take it all the way – timpani's, brass and strings, make it really over-the-top!'

The danger of this is that when someone gives Ian Curnow an inch on an orchestral arrangement, he would take a mile. Tony's requests were a license to let Ian loose and I certainly was not going to hold him back. I had no idea what Tom or the record company were going to think of this but as far as we were concerned we had encouragement by the songwriter and bandleader. Our session vocalist, Tee Green, kindly came into the P&E studio to assist us with Brian's vocals on 'Stay Another Day'; Tee had already recorded all the backing vocals in a mapping process for the band members to follow[5]. Before doing that, Tee would have also recorded a guide 'mock' lead vocal for Brian's referencing. Here are Tee's memories of that vocal session:

> It took us a couple of days to get Brian to sing the lead vocals on 'Stay Another Day' because Brian, by that time, wanted to be singing more soulfully. We were trying to get him to sing like The Beatles and he was trying to do R. Kelly [soul singer from the 1990s]. Tony's way of making things sad was to take the chords in a downward direction rather than 'up the way' because up the way was more cheerful than down the way and 'Stay Another Day' was about Tony's brother who died, he had passed away a few years before.
>
> (Green, 2015, personal interview)

The lead vocal session may well have been a couple of days. At this point in the band's career, Brian was beginning to encourage his musical genre preferences with the band and 'Stay Another Day' did not fall into his Soul/R&B vision for East 17. Ian and I assisted and coaxed Brian diplomatically through that lead vocal session and the rewards for everyone involved went beyond our hopes and expectations of that time.

The mixing process of 'Stay Another Day' was unusually difficult. The five-minute album version, fully orchestrated throughout by Ian Curnow[6], contains up to sixty audio tracks of vocals, pre-mixed to ten tracks for the final mix. There are piano and multiple keyboard sounds, and at Tony Mortimer's request, Ian and I had also recorded the Phil Collins-style programmed drums and a matching bass part onto the final 48-track analogue tapes for the mix. I was particularly keen for Tom Watkins and London Records to hear what Ian and I had recorded on 'Stay Another Day' (no one had visited the studio during the recording process) before we took our annual summer holiday with our families at the end of August 1994. Foolishly, I had only given myself one day in Strongroom Studio 2 to complete this complex mix.

Ian Curnow had spent some time helping me during the mix day with balancing the large amount of programmed orchestral parts before he departed. As I was approaching midnight on the session, it was becoming clear to me that I was not going to complete the mix as planned. My problem was that I could balance the vocals and keyboards to the orchestra, and I could balance the vocals and keyboards to the drums and bass, but I could not balance all of them together. It sounded a chaotic mess with the drums and orchestra together. Ian and I had disliked the drums from day one but had persevered with them for Tony Mortimer's sake. So, at around 3am, I said to the studio assistant 'let us lay down the version with keyboard, vocals and orchestra, send it to Tom Watkins and London Records tomorrow and I will deal with the full version including drums and bass on my return from holiday'. We recorded that version to half-inch analogue two-track and DAT (stereo Digital Audio Tape). On our return from holiday, P&E had more mixing time booked in Strongroom Studio 2 to complete 'Stay Another Day' and other tracks produced for the East 17 'Steam' album. I checked in with Tom Watkins and amongst the first things he said to me was that Tracy Bennett and everyone else at London Records *loved* 'Stay Another Day' as I had mixed it before our holiday. I said to Tom that it was not finished and I still had to add the drums and bass to the mix. His reaction was 'don't you dare touch it Phil – this is being planned as a Christmas single and everyone feels it will be a massive hit *exactly* as it is'. All that was needed was a radio edit to shorten the end and up-tempo club mixes for promotional purposes.

It was tempting to go back into Studio 2 to recall the 'Stay Another Day' mix and add those final elements but I resisted it. M People (a successful 1990s pop-dance band) were in Strongroom studio 1 across the corridor from studio 2 and asked me what I was working on and they were shocked when I said (in August) that I was mixing a Christmas single for East 17. It did seem strange to be doing that in the summer but such is the forward planning of the music industry.

Ian and I had not understood why everyone was saying 'Stay Another Day' would be a good Christmas single and it took Tom to point out that it was the tubular bells Ian and I had added to the outro making it sound similar to the bells used to celebrate the arrival of Christmas.

Much to our surprise and joy 'Stay Another Day' made Christmas #1 in the UK charts in 1994. For Ian and I, this was our biggest production success to date and we still receive producer royalties for catalogue sales, mainly due to the many Christmas compilation CD's that appear every year. Ian and I also receive PPL (Phonographic Performance Limited) performance income for our roles as musicians on the record. Tom Watkins commented:

> And like 'Stay Another Day' I feel that way exactly when you consider how that happened completely by accident. We stuck a few bells on it and called it a Christmas record and we have all done very nicely out of it ever since.
>
> (Watkins, 2014, personal interview)

Here is the full breakdown of the song arrangement.

East 17: 'Stay Another Day' (main single radio edit) 1994

Introduction x 5 bars: The first 5 bars are a typical P&E introduction with a big stereo synthesizer pad, timpani drum roll for dramatic effect and a warning of what is to come, with a complex acoustic piano (Roland sample) part from Ian Curnow. These East 17 introduction piano parts were always performed by Ian to best suit the song but had become a running joke by now (our second album with East 17) in that they were getting increasingly more difficult for Tony Mortimer to mime on television performances. By the end of the introduction, Ian and I had already introduced some of the programmed orchestra features such as pizzicato violins and strings swelling into the introduction chorus.

Introduction chorus x 8 bars: The typical pop formula of having a chorus at the beginning of the song, which has always worked commercially ever since Ian and I commonly used this methodology at PWL Studios in the 1980s. This is a full 8-bar version of the chorus that has everything in that will also be heard on chorus 1 after the first verse, including all the backing vocals, most of the orchestration and other backing instruments. By bar 5, Ian introduced the keyboard choir samples to give the second half a lift to justify the full 8 bar length.

Verse 1 x 12 bars: Features vocals, piano and minimal strings, mainly just cellos. This takes us right down to an intimate verse with Brian's vocal up front in the mix. As previously stated by Tee Green, it was hard work in the recording process getting such an intimate pop vocal coaxed out of Brian Harvey.

Backing vocalist, Tee Green, had the idea to add some subtle harmonies behind Brian on the second half of the first and third lines of each verse. This helps to give the verses contrast and allows the other boys in East 17 to join in for their television and video performances. This is always a good idea for BoyBands as it encourages the cameras to focus on multiple band members throughout the song. The first time these harmony backing vocals appear there are just two parts, by the last four bars they become three-part harmonies.

Chorus 1 x 9 bars: Enters at 1' 34", which is a little late for most pop songs but as there has been a full 8 bar chorus at the intro, this works fine. The only difference between this and the introduction chorus is an extra bar added at the end to allow a dramatic choir swell and piano riff from Ian Curnow to create a *flow* into verse 2.

Verse 2 x 9 bars: Features vocals, piano and strings again, plus the backing vocals as in verse 1. There are two major differences in verse 2 compared with verse 1. The last backing vocal is a subtle counter-line to Brian, again another idea added by Tee Green that flows into the added bar at the end of the verse to create a bigger orchestral build into chorus 2. This is the last and extended chorus as there was no middle 8 or bridge written by Tony Mortimer.

Chorus 2 (and outro chorus) x 16 bars: We go into this chorus with a timpani roll that ends on an orchestral cymbal roll and splash, Ian Curnow also added a church organ in this chorus. On the repeat of the chorus 8 bar cycle on the radio version, I edit to the equivalent point 8 bars later, where Ian and I introduced a completely new backing vocal line to give the chorus some vocal variation for the outro. This is a unison line sung by Tee Green and tracked six times to make it audible and thick enough behind the wall of vocals already present in the main chorus. The tubular bells also enter at this chorus repeat.

Ending x 4 bars: This falls out of the extremely crowded last chorus that is bursting with drama and energy as Ian Curnow built the orchestral arrangement to the maximum. Timpani, orchestral cymbals, gongs, choirs, the peeling tubular bells all finish their patterns and come to a halt like a steam train at the end of the railway track.

Conclusion

The intersubjectivity with this project stems from an unusual re-arrangement by P&E of Tony's original song demo, which had none of the extra bars that were inserted at the end of the introduction, the end of chorus 1 and at the end of the last verse. The ideology behind making extra bars work in pop arrangements is to make the listener feel that it is natural and does not disrupt the *flow* of the song. Generally extra bars are added for dramatic effect and this arrangement breakdown is a methodology example of when to insert them. The Pet Shop Boys are also past masters of the extra bar, or even half bar, in their dramatic pop productions[7]. The concept is that the listener should not notice these extra bars, until someone analyzes it, as I have done here.

One of the unrecognized roles of the music producers' tasks, in all genres of music, is the art of knowing when to change or edit song arrangements for dramatic and creative effect. Csikszentmihalyi (1997) describes where this type of creativity is by stating:

> Creativity does not happen in people's heads, but in the interaction between a person's thoughts and a sociocultural context. It is a systematic rather than an individual phenomenon.
>
> (Csikszentmihalyi, 1997, p.25)

Certainly we had good sociocultural people around us with a trusting songwriter starting the process and allowing P&E the time and space to deliver a somewhat systematic production to Tom Watkins. Tom possessed the knowledge of the cultural pop *domain* to be confident of the record's success. In early 1995, Music Week reported that East 17's 'Stay Another Day' was the third highest selling UK single for 1994. It also achieved #1 in Australia, Holland, Sweden and Austria.

7.4 'UP ALL NIGHT' (ALBUM) BY EAST 17

#7 UK Charts, October 1995

East 17 Members:

Tony Mortimer: Principle songwriter and rapper.

Brian Harvey: Lead vocalist.

Terry Caldwell: Backing vocalist and dancer.

John Hendy: Backing vocalist and dancer.

Songwriters: Tony Mortimer, Brian Harvey, Terry Caldwell, John Hendy, Dominic Hawken and Rob Kean.

Producers: Phil Harding, Ian Curnow (P&E Music) and Rob Kean.

Manager: Tom Watkins.

The commentary of 'Up All Night' by East 17 is a mixture of research, personal reflection, interviews and analysis through my experience of co-producing the album with Ian Curnow and Rob Kean. The sociology and economics around this project provide a good example of the scenario Simon Zagorski-Thomas (2014) refers to here:

> This idea of branding and the manipulation of, and response to, demand underpins a great deal of the scholarship within popular music studies. Whether considering the way businesses and musicians they contract to make their records reflect the socio-economic and socio-political landscape that surrounds them, or the way they can persuade an audience into demand for products through gatekeepers, trendsetters and marketing experts.
>
> (Zagorski-Thomas, 2014, p.227)

There was huge anticipation from the media and the audience for this third East 17 album following the UK and European success of the second album 'Steam'. The demand from the licensee labels around Europe was putting pressure on London Records and East 17's management to persuade the band to return to the recording studio and deliver. Tom Watkins and Tracy Bennett of London Records were determined to negotiate and compromise with East 17 on whatever it took to get the album underway and completed in time for the 1995 Christmas market.

The Ivor Novello UK songwriting awards take place in London each June and after a good run of hits for East 17 in 1994, culminating in the Christmas #1, 'Stay Another Day', it was no surprise that songwriter Tony Mortimer was nominated for the UK Songwriter of the Year award. Ian and I had never been to the Ivor Novello Awards before. It is a private and traditional music award ceremony that has refused to license itself for television broadcast. East 17 were in the middle of a UK tour at that time so Ian and I decided to tag along with Tony Mortimer to their evening concert in Cardiff after the awards. We had booked into the same hotel as the band in Cardiff and prepared ourselves for a long networking day at the awards ceremony, leaving the studio empty behind us, which is something that production teams are often reluctant to do. Tony won two awards and included a thank you to Ian and me in his speech after collecting the songwriter of the year award, which meant a lot to us; it is unusual for any artist at an awards ceremony to acknowledge their producers. Predictably, the media frenzy around Tony in the pressroom after the awards finished made it difficult for us to stay with Tony but thanks to his security man, Jonny Buckland, we made it to the band's MPV people carrier for the evening gig. Understandably, Tony was on a high with adrenalin and excitement but was humble about winning the top prize of the day. I think deep down he was incredibly proud to have won the awards but the rest of the band was on his mind, that evening's gig and how the guys in the band would feel about not being involved in the awards ceremony. The car journey to the concert in Cardiff was a good opportunity for Ian and me to talk with Tony about the next album and listen to some of his new song demos. The standout track on Tony's cassette was 'Thunder'; we could immediately envisage the hit production potential for this track and we were both enthused and relieved. Ian and I had already been told by Tom Watkins that the plans for the third East 17 album would be different to the first two albums and for the first time Ian and I were to produce the whole album during the summer of 1995. Tom Watkins told us that each band member would write three songs and come into the studio with us when they were ready. Tom also indicated during the album pre-production meeting that at this point of their career, East 17 had started to believe in their own (media) hype. No amount

of negotiating with Tom Watkins, his lawyers or East 17's accountants would change their minds on major decisions such as the equal
songwriting splits for the songs on 'Up All Night'.

It was clear to me that the other three members of East 17 had realized that Tony was earning more money than them. He could afford
a bigger house and having consulted their lawyers and accountants
they decided that things had to change. In my view, it would have
been more sensible for all four of them to write the twelve songs together, that way the experience of Tony Mortimer's songwriting could
have been used as a quality control throughout. At the 6 June 1995
meeting between Massive Management and P&E Music, Ian and I
suggested that all twelve songs could be a co-write, with an equal
seven-way split between the band members, Rob Kean, Ian and myself. This was dismissed by Tom Watkins as the band were adamant
that they would each go to a songwriting studio and compose three
songs with collaborators of their choosing. In my 2014 interview with
Tony Mortimer he said that the seven-way split idea was never suggested to the band but he agreed that it might have produced a better
result. Ian and I were therefore forced into a situation of having to
turn some average song demos into great sounding tracks, using our
production skills.

The recording started in July 1995, Ian and I were now contrac-
tually committed to producing a twelve-track album comprizing of
three songs written by each band member. 'Someone To Love' was
the first song of Tony Mortimer's that was recorded for this album. As
had often happened in the past, the song demo that Tony presented
on cassette was a little rough around the edges. Immediately, Ian and
I wanted to improve the arrangement and go for a big production as
we had done with Tony's songs in the past. We discussed this with
Tony over lunch around the corner from The Strongroom Studios and
he agreed to let Ian and I go ahead with the changes. I wondered at
the time if Tony had some reservations about this as he obviously
thought that his song demo was better than Ian and I did.

We re-arranged the song to achieve maximum commerciality and
as many dynamics as possible for the production. Typically Ian and I
would follow the P&E production schedule outlined in the production
appendix, recording our session backing vocalist before the band. At
this point we had been instructed by Massive Management not to use
the vocalist Trevor 'Tee' Green any longer, no reasons were given.
Here is Green's commentary about the songwriting situation for this
album and his position:

> When it came to the 'Up All Night' album suddenly you had four
> songwriters in the band and that's when it all went tits up – the
> other boys weren't in Tony's league at that time as songwriters.
> Tony relented on the second album to do something more Soul /
> R&B for Brian and wrote 'Hold My Body Tight'. I did co-write
> two songs with Brian for the third album, even though I wasn't

allowed to come into the studio and sing. On 'Ghetto Love Song' Tom took my name off the writing credits and I had to get a solicitor to sue Tom to get my money earned from the song. All of that to me was me defending myself, not attacking them. I'm still friendly with all the guys in the band and have helped them on recent recordings; I'm always available for them.

(Green, 2015, personal interview)

Ian and I hired the session vocalist, Andy Caine, who came highly recommended from a number of people including our assistant, Julian Gallagher. This became the start of a long working relationship between Andy Caine and P&E as we would also go on to do a tremendous amount of songwriting together. Tony Mortimer was present for the vocal sessions with Brian and made little comment about the rearrangement of 'Someone To Love' or the production elements. We moved onto the mix stage because Tom Watkins was asking to hear the first track completed for the album as it was earmarked to be the first single. Ian and I were very happy with the finished production, feeling we had transformed the song from fairly average sounding on Tony's demo to a commercial, dynamic production, worthy of the first single from the album and knowing that we still had the 'Thunder' song to come from Tony that we had previously heard. On the day of mixing in Strongroom studio 2, I had spent many hours creating the mix in preparation for Tom Watkins to arrive after office hours for a listening session. I had hoped he would be excited and approve the mix, allowing us to press ahead with recording the rest of the album. Tom arrived in a good mood but seemed apprehensive to me. I played him the mix of 'Someone To Love' at a fairly loud, main studio monitor volume. To us, it sounded exciting but these were Tom's words and actions: 'No babes, it started here [hand motion low down at the bottom of a hill], it went to here [hand motion upwards to the top of a hill], it went like this [hand motion in a straight line] and finished here [hand motion back down to the bottom of the hill]. It lacks any dynamics, any excitement and it was as flat as a pancake'[8]. Needless to say, I was shocked. I pointed out the dynamic lift Ian and I had created into each chorus, I pointed out the moody low key middle section that then rose dramatically into a resounding and uplifting last chorus but Tom claimed he did not hear any of that and he hated it. What is more, Tom said that Tony hated it (he was not there), Tony hated what Ian and I had done with the song arrangement and felt that we had ruined his ideas and that Ian and I had bullied him and the band into this version. Tom gave us an immediate ultimatum that was basically, scrap this version, start again from the demo, give Tony what he wants to hear, or P&E would be off the project. We were given twenty-four hours to think about the situation and the remaining sessions were cancelled until we gave Tom an answer.

It was clear to us at this point that even though Tony Mortimer and the band had seemed happy with the track, sitting in on the recording sessions, they were going back to the Massive Management offices and saying to Tom that they were not happy with the direction the track was going. Therefore, in my view, Tom had arrived for the mix session with his mind already made up from the feedback he had received from East 17, to reject it. This highlights the potential conflict of interest that can happen when the artists and the producers share the same management. I had been concerned about this potential conflict of interests since the P&E relationship with Tom Watkins began in 1992 and here we were in 1995 with that scenario becoming a reality. Keith Sawyer (2003) describes the breakdown of group creativity in these terms:

> When group flow fades away, the group usually breaks up because its members want to find new challenges elsewhere.
>
> (Sawyer, 2003, p.52)

It was clear to me this group of creative individuals were heading for some kind of break up as Sawyer describes, the question was, would it be now or could P&E hold it together to the end of this project? Once Ian and I had calmed down that evening – we had not challenged or argued with Tom at all – we absorbed his comments, took his advice (which was to capitulate and give the band and Tony what they wanted). The possibility of losing the album production, two months of well-paid work that would pay the P&E rent and bills for the rest of the year at The Strongroom and also the possibility of losing the future income of the guaranteed sale of two million albums or more (Tom's estimate) led us to conclude that P&E would be in favor of remaining on the project regardless of the creative problems. All we had to do was drop our egos, our disappointment of people not appreciating our production achievements for 'Someone To Love' and move on to produce the best of what was achievable with the song demos that East 17 would supply us to work with[9].

Two major decisions Ian and I made that evening, that would help us continue with the production of the 'Up All Night' album, were:

1. This would be the last project P&E would do with East 17 and with Tom Watkins as our manager. Ian and I would discontinue our relationship with Massive Management after this album because the potential conflict of interests that Tom had between managing East 17 and managing P&E was never more obvious due to this situation, and Ian and I now knew which side Tom was going to favor.
2. P&E would seek some programming and musical help to take the pressure from Ian throughout the album. We hired Tony King and Asha Elfenbein (ex PWL colleagues) to assist us. Tony and Asha would come into the P&E studio in the evenings at the early stages of the productions to program ambient song introductions and the rhythm elements.

The following morning Ian and I called Tom Watkins at the Massive Management office and said that we were happy to re-record 'Someone To Love', going back to Tony's demo and re-producing the track to match Tony's requirements. We also re-iterated a strong desire to carry on with the remainder of the album and were eagerly awaiting the other song demos to work from. Tom said to us that he would have to have a meeting with the band that day (already arranged) as he had cancelled that day's recording session on the band's schedule. After that meeting, Tom said he would call us back with East 17's decision. Ian and I had not anticipated this cycle of events and, needless to say, we spent a nail-biting day, sitting in the studio courtyard on our garden furniture debating where we would go from here if the band should say no and how we might fill our time over the coming weeks and months.

What was clear to us was that this was the right time to plan to move on and regain our independence from Massive Management. I am not sure we felt it at the time but there was a clear déjà vu here with what happened to us in 1991 at PWL, when the production deal for us to produce Take That fell through due to Pete Waterman's decision not to allow the PWL credit to be removed from the contract and therefore the Take That record. That triggered a clear reaction from us that it was time to leave PWL because it was becoming detrimental to our work and our future careers to stay there. Now the same was happening here in 1995. P&E's best work with Tom Watkins and East 17 was probably behind us and Tom Watkins was clearly not prepared to fight the P&E corner with his own artists, including East 17. So this was another 'light bulb' moment, except that unlike 1991 when we stayed at PWL for almost another year, here Ian and I knew it was just two months and once the word got out that we had left Massive management, work and offers would come to P&E as a songwriting and production team. Ian and I still had our publishing deal with BMG Publishing and support from Paul Curran and Mike Sefton; they had been delighted with the run of money-spinning hits P&E had delivered in 1994 and 1995 with Deuce and East 17.

Finally, later that day, East 17 arrived at our studio and apologized. Slightly bewildered by events, we were back in business. The remainder of the recording and mixing sessions for the project were fairly unspectacular. Ian and I worked very hard throughout July and August to record the chosen twelve songs for this album. We created a concept of linking many of the tracks with ambient-style introductions and outros to give some added credibility and interest to some of the songs, especially those written by John and Terry. These sections were the main programing parts that Ian and I delegated to Tony King and Asha Elfenbein who would come into the P&E studio overnight so that Ian and I could remain on schedule to deliver the album.

'Someone To Love' did eventually become a single taken from the 'Up All Night' album. It was not a big hit, reaching #16 in the UK singles chart. It is likely that Ian and I kept some of the elements, such as backing vocals from the version that Tom Watkins had rejected. The major aspect was going back to the original song arrangement and instrumentation base and then re-recording Brian Harvey's lead vocal, with Tony Mortimer present to approve the process and arrangement. Thankfully, the first single released from the album, 'Thunder', was a top 10 hit all over Europe. East 17 delivered a tremendous live performance of 'Thunder' on the MTV Awards later in 1995 and it pushed the album sales well beyond the two million that Tom Watkins had predicted. Whether the original P&E version of 'Someone To Love' would have made top 10 instead of top 20, we will never know – it has never been released or heard to this day. The main point was that, to the extent that management, band and label needed to be, all were happy with the P&E production on the 'Up All Night' album.

Unlike the previous East 17 albums, Ian and I did not see London Records head of A&R, Tracy Bennett, once during the whole album studio process. I think he was aware of the diplomacy and problems going on within the band and London Records, together with their European affiliates were delighted to get a third album out of East 17. Beyond the further singles that were released from the 'Up All Night' album, the only other significant recording by East 17 in 1996 was the single 'If You Ever' with Gabrielle, which was in the Soul/R&B genre that Brian Harvey had been pushing the band to go towards. East 17 would split soon after that when Brian Harvey unfortunately made the mistake of talking on live UK radio about the drug, ecstasy, and its merits. That became front-page news and allowed Tom Watkins to remove Brian from the band. Some years later, East 17 would reform as a three-piece without Tony Mortimer and release a Soul/R&B style album that Ian and I collaborated on for a couple of songs.

By the end of the September 1995 P&E had completed the last pieces of work for the East 17 and Deuce albums. Ian and I arranged a meeting with Tom Watkins and he thought that we were coming in to discuss his new signings, such as boy duo 'North & South'. Ian and I knew we were going in to have a tricky meeting, to say that we had decided to move on and seek new management and new production and songwriting projects of our own. Much the same as our meeting with Pete Waterman in 1992 when we departed from PWL, Tom Watkins was professional and understanding and he in no way tried to dissuade us as he could see that we had considered this carefully before the meeting and we parted amicably with a handshake, wishing each other well for the future.

Conclusion

This commentary highlights that the *flow* of creative collaboration had rescinded between the *agents* in this particular cultural *domain*. Ian and I had kept the atmosphere in the creative *field* (the P&E Recording Studio) as positive as we possibly could. P&E had reverted to the most extreme example of a creative team working to a system akin to a traditional service industry, such as catering, bearing in mind the whole time that commerce had taken priority over creativity due to the previous successes in the *domain* of this same set of *agents*.

By November 1995, the first single, 'Thunder', from East 17's album, 'Up All Night', peaked at #4 in the UK charts and #3 in the German charts. The Music Week review for the 'Up All Night' album was critical of the songs and the performances. The album was released on 25 November 1995 and entered UK album charts at #7. By December 1995 the East 17 overseas album sales were reported in Music Week as slightly fewer than two million. The album was still doing well in early 1996 with 'Thunder' reaching top 10 across most of Europe. The second single from the album was 'Do U Still', a more Soul/R&B sound and it peaked at #7 in the UK charts in February 1996.

7.5 'WORDS' (SINGLE) BY BOYZONE

#1 UK Charts, October 1996

Boyzone Members:

Ronan Keating: Lead vocalist.

Stephen Gateley: Lead and backing vocalist.

Keith Duffy: Backing vocalist.

Shane Lynch: Backing vocalist and dancer.

Michael Graham: Backing vocalist and occasional lead vocalist.

Songwriters: Robin, Barry and Maurice Gibb.

Producers: Phil Harding and Ian Curnow (P&E Music).

Manager: Louis Walsh.

This commentary of 'Words' by Boyzone is a personal reflection and analysis of the seventh single release by Boyzone; it was their first UK #1. In the process of reflective analysis it can be difficult to explain the 'lived experience' of music production projects. I am going to present a systematic examination of how a pop production team in the 1990s would prepare a cover song arrangement, looking closely at the 1960s original version by the Bee Gees and creating an interpretation that would be culturally suitable for the 1990s British BoyBand marketplace. The explosion of successful BoyBands on the 1990s British pop market was nothing short of a phenomenon and having been major *actors* in the creative *field* and *domain* that surrounded successful BoyBand, East 17, prior to this project, Ian Curnow and I had the confidence that Boyzone's manager, Louis Walsh, would allow us to practice the technique of 'free imaginative variation' (Husserl, citied in Smith, Flowers & Larkin, 2009, p.14) in our approach to re-arranging this classic 1960s song.

Ian and I had no idea how long our run of projects with pop manager, Louis Walsh, would last but in fact it was surprisingly short. Our relationship with Louis started when we had parted company with Tom Watkins in October 1995. P&E entered into a good *flow* of production and songwriting work for Louis Walsh in early 1996 and it peaked by the autumn with Boyzone's 'Words' reaching #1 in the UK singles chart, the second #1 single for P&E in consecutive years following the success of 'Stay Another Day' with East 17.

Songwriters and artists, the Bee Gees, originally composed 'Words', achieving a UK #8 hit in 1967. It was a song favored by Louis Walsh, who loved the whole Bee Gees catalogue and Louis was regularly asking Ian and I if their songs would work for a modern 1990s cover version for one of his acts, generally the BoyBands. Louis felt strongly that the song would work for Ronan Keating's voice and a Boyzone sound boosted by a P&E orchestrated sound that had become a trademark for us thanks to Ian's skill as an arranger and our large bank of orchestral MIDI sounds and samples[10]. That type of orchestrated dramatic sound was something that Louis liked about the P&E productions with East 17. Ian and I looked at the song arrangement from the Bee Gees original and felt that we could significantly update and improve it to commercialize the song for the 1990s BoyBand genre. As a case study, it is interesting to compare the differences between the two song arrangements.

'Words': The Bee Gees version: 1967 piano ballad 80 BPM approximately

Introduction x 4 bars: An atmospheric piano and guitar build up.

Verse 1 x 8 bars: Vocals/piano/guitar and bass.

Bridge 1 x 4 bars: Vocals/piano/guitar and bass. The chords begin to build.

Verse link x 4 bars: This goes down and back to the verse chords to set up a second verse. Ian and I felt that this was not very commercial for the 1990s BoyBand genre and the BBC Radio 1 and 2 formats of that time. We decided to cut these 4 bars, double the bridge and take the listener straight into the chorus. This section is consequently never used on the Boyzone version.

Verse 2 x 8 bars: Drums enter here. Orchestra joins around bar 5 of this verse to help lift the track into the second bridge.

Bridge 2 x 4 bars: Builds into the chorus.

Chorus 1 x 4 bars: Enters at 1'35", that's almost the halfway point of the song and only goes around once. This goes against the philosophy that Ian and I had experienced whilst working for Pete Waterman at PWL in the 1980s. Waterman would always insist that the first chorus must enter a song arrangement by one minute maximum, especially if there had been no chorus at the introduction.

Middle 8 x 8 bars: Vocal 'la-las' over the verse chords.

Bridge 3 x 4 bars: – Second bridge repeated.

Chorus 2 x 12 bars: This time the chorus repeats three times, with the third repeat coming down to just guitar and vocal.

Ending x 5 bars: Very similar, isolated piano sequence of the introduction chords.

'Words': The Boyzone version: 1996 piano ballad 82 BPM approximately

Introduction x 8 bars: The first 4 bars contain a typical P&E solo piano with some orchestra fading in played and arranged by Ian Curnow, very similar to some previous East 17 introductions that Ian and I had produced on their second album 'Steam'. By bar 5 the sequence is similar to the Bee Gees introduction.

Verse 1 x 8 bars: Vocals, piano and strings. The record company specifically requested that the first verse focus just on Ronan Keating's vocals with minimal backing.

Bridge 1 x 8 bars: As explained on the Bee Gees version, Ian and I felt it was more commercial to keep the bridge building for the momentum of the song and to go straight into the chorus. We combined the first and second bridges (4 bars each) of the Bee Gees version to construct our 8 bar bridge. We also added backing vocals and harmonies here to be sure that the other Boyzone members could become involved with the track during television and video performances.

Chorus 1 x 4 bars: Enters at 1' 10", almost 10 seconds later than the Pete Waterman formula of reaching the first chorus by 1 minute.

Verse 2 x 8 bars: Vocals, piano, strings and percussion. The orchestral arrangement builds throughout this verse and the percussion indicates that there is more rhythm to come.

Bridge 2 x 4 bars: Drums enter here with a big fill and build into the chorus, these 4 bars are lyrically the same as the third Bee Gees bridge.

Chorus 2 x 8 bars: Ian and I add the third harmony above Ronan on the repeat (fifth bar) of this chorus to keep it building.

Middle 8 x 8 bars: Vocal 'da-da' over the verse chords with the full backing still in, including drums. At this point of the record, on the Boyzone BBC1 Top of the Pops (TOTP) appearances, all five boys would rise from their seated positions on high stools and walk towards the audience and the cameras, causing loud cheers from the TOTP audience. That sounds dull now, but it was exciting the first time one saw it happen and it was a dynamic visual lift for the record.

Bridge 3 x 8 bars: Duplication of our first double bridge but with the full backing (drums & bass) still in and an expanded orchestral arrangement and a large piano flurry at the end to *flow* into the last choruses. The track was now becoming the typical 'P&E big production sound' that Louis had requested from Ian and I.

Chorus 3 x 12 bars: The third vocal harmony is present immediately and as per the Bee Gees version, Ian and I strip out everything by the third repeat of the chorus, leaving just Ronan and the piano, plus the orchestra, much the same as the introduction. The hope being that we have taken the listener on a journey throughout the track, from a subtle piano introduction that builds and builds to such a peak that the only place to go by the last chorus repeat is back down to where we started.

Ending x 4 bars: Very similar isolated piano sequence to the introduction.

These examples, comparing and changing song arrangements, is the type of preparation that producers in all genres of music will go through before recording sessions. The reviewing of arrangements will then carry on throughout the recording process and the final decisions will be made during mixing and post production editing, especially in the classical world, where multiple post production edits will be performed by the producers and engineers. Sometimes even the mastering engineers will be called upon to do this if the producer or artist are still changing their minds at that point. Ian and I had delivered a finished version of 'Words' by Boyzone before our 1996 summer break and I was slightly disappointed to hear that we were asked to recall the mix on our return, to make some adjustments for the Polydor A&R department and Louis Walsh. Those mix adjustments were fairly minor, achieved in half a day at Strongroom Studio 2 with a small amount of reprogramming from Ian. On the importance of production decisions on song arrangements, Richard James Burgess (2013) states:

> Given a strong song and artist, influence over the arrangement is one of the most powerful tools a producer has and can spell the difference between success and failure. If you think of a song being a museum or art gallery full of interesting objects, the structure works like a curated tour.
>
> (Burgess, 2013, pp.95–96)

I have given a curated tour of the P&E song arrangement decisions in this commentary of the song 'Words'. Ian and I felt that we had achieved a significant commercial improvement on the song. Neither Louis Walsh nor the record company questioned our song arrangement decisions; the same goes for Ronan Keating and the other Boyzone members. The publishers of the song made no comment to us, one can only assume that they and the Bee Gees were delighted that we had assisted in taking the song to its highest ever UK chart position. Ian and I naively thought that we might receive a comment or message of congratulations from the Bee Gees themselves, saying that they appreciated our achievements creatively and commercially for their song, with a significant new arrangement. Nothing was forthcoming to the best of my knowledge, to Louis, the publisher or P&E Music.

The success of the Boyzone record did spark a 'congratulations and thank you' Sunday night phone call from Louis Walsh and Ronan Keating to my home in Suffolk, after the chart position had been announced on Radio 1. This was a kind and unexpected gesture that I shared with Ian as soon as the conversation with Louis and Ronan finished. Not many managers or artists in the world of pop music would take the time to show this appreciation to their producers.

Conclusion

Irish BoyBand, Boyzone, had six UK #1 chart singles; 'Words' was their first. It was the tenth biggest selling BoyBand single of the 1990s and the sixteenth biggest selling single of 1996. It was a chance for Ian Curnow and I to prove to the pop industry that our BoyBand producer success was not restricted to just East 17 or being managed by Tom Watkins. Whilst we were under Tom's management he flatly refused to allow Boyzone's manager, Louis Walsh, to speak to Ian and I directly, the same rules also applied to Simon Cowell. Therefore, Louis was delighted to hear that Ian and I had parted company from Tom Watkins in October 1995 and this record was the result of our most successful collaboration.

As busy music producers 'we are constantly caught up, unselfconsciously, in the everyday *flow* of experience' (Smith, Flowers & Larkin, 2009, p.2) and a *flow* of events as I have described in this commentary, followed by an appreciation of thanks in the Sunday night phone call that allows us to stop for a moment in time, reflect and even comprehend a 'lived experience'.

7.6 'BOYZ UNLIMITED' – CHANNEL 4 TELEVISION SHOW

Production Company: Hat Trick Productions.

Broadcast: Monday 8 February 1999 (the first of 6 episodes)

The Main Cast:

James Corden: Gareth Jones, the songwriter and the 'the fat one'.

Lee Williams: Scott Latisier (Giles Hornchurch), 'the good looking one' who cannot sing.

Adam Sinclair: Jason Jackson, the dark haired 'tough one', Scottish.

Billy Worth: Nicky Vickery, the blond 'cute' little one who can sing.

Frank Harper: Nigel Gracey, the Manager.

Jo Whiley (BBC Radio 2 DJ): The Narrator.

Written by Richard Osman and David Walliams.

Devised by Matt Lucas, Richard Osman and David Walliams.

Produced by Richard Osman.

Episodes 1–6 can be viewed at: http://www.channel4.com/programmes/boyz-unlimited

This commentary is a reflection on the P&E music contribution to the 1999 television show 'Boyz Unlimited', broadcast on Channel 4. Television production company, Hat Trick Productions, were looking for a songwriting and production team with BoyBand success and experience to create the music for the show. This would include original BoyBand songs, typical BoyBand covers and incidental music. The television department at our publishers, BMG Music, had suggested P&E Music to Hat Trick Productions and Ian and I had to compete with other music producers and songwriters to secure the project. That was achieved through a continued set of meetings, which seems to be a favorite pastime of television production companies. Ian and I also presented a specially-written demo song for the show to highlight P&E's commitment and suitability to the project. It has been reported that the co-creators of the show were Matt Lucas and David Walliams of television's 'Little Britain' fame, although Ian and I never met them on set or in the studio. A young James Cordon was one of the BoyBand members in the show (playing 'the fat one'); it was cynically exaggerated to make him look like an early version of Take That's Gary Barlow. Our backing vocalist and co-songwriter, Andy Caine, spent many hours at the P&E Music studio, writing and recording for this project in March 1998. Andy recorded all of the guide lead vocals and harmony vocal parts for each of the four band members who were to become Boyz Unlimited[11]. Ian Curnow, still suffering from the after-effects of pneumonia, had bravely battled through March and early April for the sake of this project. We managed in this time to write the main theme song 'Stronger Everyday' for the television show and that secured the deal for P&E with Hat Trick Productions and the Boyz Unlimited project. The show was broadcast in early 1999. The actor, Frank Harper, who played the part of the band's manager, was later to star in many British movies, such as 'Lock, Stock and Two Smoking Barrels'. This project proved lucrative for P&E Music once it reached full pace in late April and throughout May 1998 and is a good example of a production team 'going the extra mile' to pull in a project under intense competition. Ian and I gently persuaded the Hat Trick television production team that we were the right people for the project.

As a slice of television pop culture, this was a first really, in as much as it was a music *mocumentary* (notably, 'Brian Pern – A Life In Rock' broadcast on BBC Four in February 2014 has been very popular and is now in its fourth season). Boyz Unlimited was cynical of the BoyBand culture; from the audition process through to the success and downfall. Its humor relied on the mock documentary style that we later saw with 'Little Britain' (also conceived, written by and starring Matt Lucas and David Walliams) and Ricky Gervais's 'The Office'. There are many camera shots throughout where after stating something ironic, the band's manager, Nigel, or a band member, gives a knowing look to the camera, a very British style of humor that has mainly restricted the show to British television viewing only. Many ironic situations are created and played out once the band has been formed from the politically incorrect and racist audition process. The band visit Abbey Road Studios (in Chelmsford, Essex, ironically) to record the first single, a cover of 'Say A Little Prayer'. BBC Radio 2 DJ, Jo Whiley, adds a convincing narration throughout, adding even more irony to the program. At the point in the recording studio where it is shown that none of the boys can sing very well, our regular session backing vocalist, Andy Caine, is shown singing the lead vocals whilst the boys are around a microphone just performing hand-claps.

The Boyz Unlimited television show music was recorded in 1998 and the show began to air in February 1999 on Channel 4. Press reporter, Andrew Smith,[12] put an interesting spin in his review that talked about the BoyBand genre generally. He credited Ian Curnow as the show's joint musical director with the show's creator Richard Osman[13]. Ironically, in between our recording for the show in 1998 and its Channel 4 broadcast in 1999, BoyBand 911 had also decided to record a version of the Doctor Hook classic, 'A Little Bit More'. We had recorded this for the show because Richard Osman thought it was such an awful song for a BoyBand to cover, that no one would actually record and release it seriously. The 911 version charted at #1 on 17 January 1999 in the UK, just before Boyz Unlimited was broadcast on Channel 4. As a known and respected rock music critic, here is the way Andrew Smith describes the BoyBand genre and the people around them:

> Most of the stuff [music] is abominable, throwaway pop. Faceless songwriting teams, who provide hits for bands such as Boyzone and Backstreet Boys at best ape, at worst parody. There is more to life than treacle ballads and damp down disco, but not if you're a BoyBand. Managers are clueless things; record companies are run by cocaine-addled airheads. The would-be stars [BoyBand members] are likeably deluded ingénues who are quickly denuded of an illusion and/or principles in the rush to fame and fortune.
>
> (Smith, 25 January 1999: *The Guardian*)

Smith goes on to state that classic pop songwriters, such as Lennon and McCartney, Neil Diamond, Carole King and Gerry Goffin, forged the songwriting formulas and he cites Richard Osman's depiction of managers, record companies and BoyBand members as 'reasonably accurate'. Smith further criticizes that the UK chart round up for 1998's biggest sellers is full of pop drivel, for example, The Corrs, Boyzone, Celine Dion, Steps, B*Witched, Five and Billie. He reports that when Richard Osman was trying to sell the show to Hat Trick Productions and Channel 4 in 1998, their biggest concern was that BoyBands would no longer be popular in 1999 when the television program would be broadcast. Richard persuaded Hat Trick Productions that BoyBands would still be massive but the end was in sight. East 17's ex-manager, Tom Watkins (2016), cites Paul Morley's[14] quote 'once something is understood, it stops being a source of energy' in his book, *Let's Make Lots of Money: Secrets of a Rich, Fat,*

Gay, Lucky Bastard (Watkins, 2016, p.148) and I think we could relate that to what Boyz Unlimited did to the BoyBand phenomenon of the 1990s. It went into swift decline by the end of the decade with Westlife being one of the few 1990s BoyBands to retain success into the 2000s. My personal view was that the genre and popularity of BoyBands was fading and that was one of my motivations for P&E to be involved with the Boyz Unlimited show. Louis Walsh, Boyzone, 911 and Steve Gilmore had all moved on from working with P&E Music and Ian and I were being offered similar European BoyBands, such as Get Ready from Belgium and CITA (Caught In The Act) from Germany, and even these BoyBands were dropped by their record companies by the end of 1999. So I did not feel uncomfortable about being involved in a satirical television show, making fun of the BoyBand genre and the music industry circus that surrounded it. Music journalist Steve Jelbert of *The Independent* newspaper wrote:

> The time has come to honor the genre [BoyBands] with a comedy series. It tells the story of a fictitious gang of ingénues unleashed on an over-crowded market, in the show the band have a couple of hit singles but are quickly seen to fail and fall apart across the six part series.
>
> (Jelbert, 25 January 1999: *The Independent*)

Jelbert goes on to describe Boyz Unlimited as a 'popumentary'. *The Independent* also reported that Larry Parnes, the manager of Billy Fury, was the inspiration for the manager in the Boyz Unlimited show. Later, for a short run in 1999, came the London West End Theatre presentation 'BoyBand the Musical', produced by theatre impresario Adam Spiegel. The show was another parody of the genre and it was criticized by the media and public for being too specialist a subject for the London stage.

The Boyz Unlimited project contained four fully produced singles by P&E for the main body of the show plus 90% of the incidental music featured throughout the series. Ian and I visited the film set at one point to get a feel of how the production was progressing. It was useful and informative for us to meet the other actors, chat to them about the script and take on board their views about the potential of the program. All of the actors seemed delighted to have the opportunity to be involved in what they described as a high-quality script and reasonably budgeted television production. To get the four main tracks sounding convincing for the show and typical of the 1990s BoyBand music genre, Ian and I treated the four main actors as a real BoyBand. During the recording sessions each band member was given a chance to sing lead vocal parts throughout the songs and all of them were given harmony parts and coached by Andy Caine in the studio. By the time P&E had finished recording the singles and the television producers and writers had storyboarded four pop videos for each of the four songs throughout the show, it all looked very realistic and convincing to us. This made it all the more hilarious when these clips were featured in the show, dance routines were choreographed for the band and were deliberately exaggerated to look bad in order to gain more laughs. Unfortunately, the show has never been released on DVD so the Boyz Unlimited series remains unavailable for many to watch. The viewing figures when the show was broadcast in 1999 were very good and everyone involved was delighted with the results and contributions that the P&E team made to the music. This was probably P&E's biggest success in an attempt to move into the television and film world in 1998 and 1999. Channel 4 clearly promoted Boyz Unlimited as a comedy because the broadcast slot was at 9.30pm, between Friends (9.00pm) and Frasier (10.00pm).

Conclusion

There are good examples of group creativity throughout this project commentary and for P&E Music it was unusual to be collaborating so closely with other creative individuals from the television and film industry. This helped to inspire P&E's music creativity and enthusiasm for the project. At times in the recording studio, it felt as though Ian and I were working with a real BoyBand. This was aided by the fact that actor, Billy Worth, had been in a BoyBand called The One, which enabled Billy to encourage the other actors to perform in the recording studio beyond the capacity that one might expect of young actors.

Main Songs Featured in the Boyz Unlimited Television Broadcast

D.I.S.C.O.: Cover of the 1979 hit by French act, Ottawan.

I Say A Little Prayer: Cover of the 1967 hit by Dionne Warwick.

A Little Bit More: Cover of the 1976 hit by Dr Hook.

Stronger Everyday: Original song written by Harding, Curnow and Caine for Boyz Unlimited.

7.7 EXTRA PRODUCTION COMMENTARY: 'IT'S ALRIGHT' (SINGLE) BY EAST 17

#3 UK Single Charts, 1993

East 17 Members:

Tony Mortimer: Principle songwriter and rapper.

Brian Harvey: Lead vocalist.

Terry Caldwell: Backing vocalist and dancer.

John Hendy: Backing vocalist and dancer.

Songwriter: Tony Mortimer.

Producers: Phil Harding and Ian Curnow.

Manager: Tom Watkins.

The production appendix commentary of the P&E production 'It's Alright' by East 17 describes a creative and cultural process between manager, Tom Watkins, producers, Phil Harding and Ian Curnow, the artists, East 17, and their record company, London Records.

To release a fourth single before an album re-launch that would have this track featured was unusual in itself and adventurous as most in the 1990s would release only two or three singles from an album. Ian and I assumed that we would be moving onto the second East 17 album by the middle of 1993, following three hit singles from the successful East 17 debut album, 'Walthamstow'. London Records and Tom Watkins felt that the band were not ready to start the follow-up album and that the first album could remain in the UK album charts for an extended period into 1993 if another hit single was produced and included with a new pressing and extra promotion. It felt to us that the pressure was on to deliver a high charting hit single. Tom Watkins was not joking when he stated to Ian and I at the start of this project 'It HAS to be brilliant, bold, different and brave'. In the studio, Watkins spent a long time driving Ian and I to take it further than we probably would have without his guidance. In the book, *Let's Make Lots Of Money: The Secrets of a Rich, Fat, Gay, Lucky Bastard* (Watkins, 2016) Watkins states:

> My favourite East 17 song was 'It's Alright'. Opening like a piano-led gospel ballad, it quickly shifted gears into a fist-pumping club-bound banger. I had insisted that the tune got speeded up to turn the big, churchy sound into pounding chant. I couldn't imagine those weedy Take That boys coming up with anything as inventive or surging as this.

> (Watkins, 2016, p.306)

As Watkins states, the introduction's flamboyant piano style from Ian Curnow is almost Liberace in its form[15]. When the drums and the remainder of the instruments enter, we are hit with a typical BoyBand chant and heavy electric guitars. Watkins drove the guitar sounds and the backing vocal sounds. The distorted electric guitar sounds are actually Ian Curnow using a set of Roland keyboard sounds that were routed through a rack amplifier made by Rockman and over-driven by our DBX160 compressor. Ian mastered the art of playing guitar-style chords, shapes and even lead guitar licks on the keyboard and creating realistic electric guitar sounds. As aforementioned, Ian and I nicknamed this sound 'Eddie' after the famous guitarist, Eddie Van Halen, lead member of the successful American rock band, Van Halen. Processed big drum sounds and aggressive keyboards through-out the up-tempo section of 'It's Alright' completed the picture of a hit single that showed what was to come in the future from East 17 – more class, fun and sophistication in the up-tempo and down-tempo sections. When 'It's Alright' was released in November 1993, it shocked many industry and media people, pleased the East 17 fans and laid down a clear marker to Take That and other UK BoyBands as a quality song, production and performance benchmark. In other words, 'TOP THAT – IF YOU CAN' (Watkins, 1993). Ian and I were delighted with the final production of 'It's Alright', pleased with its chart performance and very hopeful that we would be asked to produce the bulk of the forthcoming East 17 second album that would surely start soon. We were therefore a little surprised to hear that yet another additional track had been recorded for 'Walthamstow', produced by The Groove Corporation and Mykael S. Riley. Ian and I reluctantly agreed to remix this track (a cover of Pet Shop Boys' 'West End Girls'), which was only used on the 12-inch single B-sides. Ian and I waited patiently to hear what would happen on the second album, which when it finally started was to prove a shock and disappointment for us. Producers, Richard Stannard and Matt Rowe, were commissioned to co-write and produce the first two tracks towards the second East 17 album 'Steam'.

NOTES

1. Tom Watkins became P&E's manager as well at around the time that 'House of Love' was released.
2. Another Germany-based music sequencer company, originally known as C-Lab.
3. Nicknamed 'Eddie' (after rock guitarist Eddie Van Halen). See the Appendix.
4. The mixing process of 'Deep' is briefly discussed in Chapter 6.
5. The P&E Music typical BoyBand vocal production methodology is described in Chapter 5.
6. Programmed using Akai samples, Roland sounds, Proteus sounds and Korg sounds. See more in the Appendix.
7. I edited the original mix of 'Opportunities: Let's Make Lots of Money' for Pet Shop Boys at Abbey Road Studios and was shocked to discover extra bars and even half-bars used to dramatize the arrangement.
8. These were Tom's words from my memory of the day.
9. Making productions from songwriting demos was still a common practice in the 1990s. This has gradually dissipated to become a start point for a production today, where the demo files are passed over to the production team.
10. A full list of the P&E Studio MIDI equipment can be found in the Appendix.
11. The same BoyBand vocal production system as described for East 17 in Chapter 5.
12. Smith wrote the rock music column page of UK newspaper, *The Guardian*, during the 1990s.
13. Richard Osman is the co-creator of the BBC 1 television program, 'Pointless'.
14. Paul Morley: Former New Musical Express music journalist who joined producer Trevor Horn's record label ZTT Records in the 1980s to co-create Frankie Goes To Hollywood.
15. Liberace was a flamboyant American pianist and entertainer; his piano style was as 'over-the-top' as his image.

8

Conclusions and Theories

8.1 INTRODUCTION

In their publication, *The Art of Record Production: An Introductory Reader for a New Academic Field*, Frith and Zagorski-Thomas (2012) refer to the new academic study of contemporary and popular music production that started in the UK with the commencement of university courses, such as the Tonmeister course at the University of Surrey in Guildford in late 1970s. Music production itself is not new and sociologist, Theodor Adorno, makes reference to the term in his essay *On Popular Music* (Adorno, 1941). But just as techniques and production have changed, so has its status as a subject of academic enquiry. In part, this book represents the shifts in music production and academic thinking, which has allowed me to reflect on my professional production work during the 1990s. As such, my hope is that my professional practice will inform and contribute to academic debate in this field.

The outputs of my PhD led me to the two main theories featured in this book: A 'Service Model for (Pop Music) Creativity and Commerce' and 'Group Creativity and Human Interaction'. As a creative industry practitioner of more than forty years, I consider my commercially successful period throughout the 1990s as my peak *flow* period as an artist. I report on Mike Stock's view[1] that he and Matt Aitken were co-artists with the singers during their successful pop music period throughout the 1980s. Ian Curnow and I viewed our collaborations with the manufactured pop and BoyBands throughout the 1990s in the same way. Ian and I were the artists providing the foundation of recorded music that the named artists would sing over and then promote to the media and public. Pop music production can easily be overlooked in the world of academia and compared with other music genres, such as rock, dance, classical and jazz (Frith, 1996), pop is often relegated to the field of sociology or cultural studies rather than taken as an example of good music production or performance. I would suggest that there is 'a gap in the network of knowledge' (Csikszentmihalyi, 1997) for pop music production that is rarely addressed at academic conferences, such as The Art of Record Production (ARP), or anecdotally during lectures at universities and colleges worldwide. Frith offers this view in *The Art of Record Production: An Introductory Reader for a New Academic Field* (Frith & Zagorski-Thomas, 2012):

> Record Producers are seen both more significant for rock as an art form than producers in jazz, folk or classical music, but less important for rock as a cultural project than producers in pop or dance music.
>
> (Frith & Zagorski-Thomas, 2012, p.221)

Frith is hinting here that producers in the pop and dance genres have a significantly different role to music producers in other music genres. I would agree with this and much of my commentary throughout this book indicates a somewhat unique position for the role of pop producers. Pop music production has changed enormously since the 1970s and 1980s. The technology developed since then means that anyone can now make a high quality recording at home and release it to the public through social media in a short space of time. However, the analogue technologies and techniques of record making from the 1970s are still with us and inform many of the processes used today. It is important for anyone wishing to enter the field in the twenty-first century to possess the contextual background of the pop production industry as it has progressed since the 1960s and 1970s, as well as the current practices and methodologies. Mihaly Csikszentmihalyi (1997) emphasizes that the success of creative literature writers largely depends upon:

> Immersing themselves in the domain of literature. They read avidly; they took sides among writers, they memorized the work they liked – in short, they internalized as much as they could from what they considered the best work of previous writers. In this sense, they themselves became the forward-moving edge of cultural evolution.
>
> (Csikszentmihalyi, 1997, p.262)

This immersion into the domains and methodologies of previous pop records is directly attributable to the success of pop and BoyBand producers during the 1990s, such as Ian Curnow and myself. By the middle of that decade, Scandinavian production teams, such as Cheiron Productions in Sweden, who had clearly studied the manufactured pop phenomenon in the UK and America throughout the late 1980s and early 1990s, began to make an impact on the worldwide pop marketplace with acts such as The Backstreet Boys, N-Sync and Britney Spears. At PWL Studios throughout the 1980s, Pete Waterman would be the source of other relevant pop records in the domain for the SAW production team to use as references. Every weekend, Waterman would receive pre-release promotional copies from record companies to play on his Saturday morning radio show in Liverpool. Waterman would then spend Saturday afternoon observing what the local dance record shop was playing and selling to the local club DJs. Anything that stood out to him as relevant for referencing for SAW songwriting and productions would then be presented to the team on Monday mornings. I soon realized that I also needed to do my own research and source records that were current and relevant to the projects that Ian Curnow and I were remixing and producing at PWL Studios during the 1980s. My observations of the PWL 'Hit Factory'[2] success and Pete Waterman's methodologies were useful training for the successful P&E Music songwriting and production period that followed throughout the 1990s. Ian and I achieved a string of hit records with East 17, Boyzone, Deuce and others, largely thanks to our internalizing of what we considered the best work of SAW and others throughout the 1980s.

However, over and above this is the requirement to understand the skills, attributes and the *being* of the pop music producer. People in the music industry may not agree with this, but in my view, the music industry is a 'service' industry. If there is an area or genre of music production where one can say that this *service industry* view is prevalent, it is in pop music production. Quite possibly this pop music 'service industry' has existed since the 1950s and 1960s but largely during the artistically indulgent music of the 1970s it took a back seat, whilst budgets for rock and concept albums were at their highest ever level and popularity. The music critics and fans of what was considered 'serious music', such as progressive rock acts like Led Zeppelin, sneered upon pop music of the 1970s by pop acts, such as ABBA. As long ago as 1941, Adorno (1941, 2002, p.437) stated that there were two spheres of 'popular music, characterized by its difference from serious music'. Adorno describes those differences using the origins of classical (serious) music from Europe and popular music arising from America. So the stigma of popular music not being serious goes back as far as the 1940s but was taken to another level in the 1970s when classical music was not even in the discussion. Adorno's comments highlight an historical trend to denigrate popular music at the expense of another type of music. For Adorno it was classical music, in the 1970s the rock album being characterized as 'serious music' had replaced classical music. It is difficult to sustain these arguments because poplar music has enormous value to a wide range of the population. Value can be expressed economically, socially, culturally or artistically and the modern day pop producer has made a contribution to all of these areas as a 'service agent' to record companies, artists and managers (albeit a creative agent as well). The pop producer navigates a specific role between the commercial and creative demands of the industry.

More often than not today, the producer is not one individual but a production team consisting of anything from two to four members.

That team will generally consist of one or more musicians, with at least one keyboard player and programer, able to navigate the current music sequencer packages, such as Pro Tools, Logic Pro or Cubase. There will often be an engineer, capable of not only being able to operate and edit those same music sequencers, but also with a good knowledge of recording techniques, microphones and most importantly, creative mixing talent and the ability to 'excite' the room (the artist and the other production team members) with a good working balance for overdub recording and ultimately a final mix balance that completes 'the picture' that everyone has been working towards. In my view, there will need to be a 'team-leader': An entrepreneur with good organizational skills, good administration skills, excellent communication skills and most importantly of all, the ability to make the projects run smoothly from start to finish. That includes taking the pressure applied by the client (generally the record label or artist manager) away from the creative members of the team, so that they can concentrate on music creativity only. The team-leader also needs to be an excellent diplomat, which is a typical skill incorporated by most successful music producers. Therefore the 'team-leader' will be the person with the most novelty in the cultural domain (Csikszentmihalyi, 1997), the instigator of the project to the team and will also be the final 'filter' of the product from the field back to the domain.

Increasingly, pop production is achieved remotely, with producers and musicians in their own recording studios, working online, exchanging files and ideas by email or large file download sites, such as Dropbox. Generally, the production team will begin by uploading files of a basic backing track, with guide vocals, to a set of session musicians, keyboardist, guitarist, bassist, drummer, and so on. Then all of those musicians will be required to have their own home studio set up where they can perform and record, in their own time, the requested overdubs that the producer has asked for, plus any other ideas they wish to present to the production team for consideration.

It is entirely possible that one person can acquire all of these skills and abilities and be able to produce a superb and successful record, but these individuals are rare in the twenty-first century. I do not believe that one person should seek to master all of these skills and abilities; my view is that it is preferable to excel at one or two and collaborate with or delegate the other jobs to partners, team members and session musicians. Often, an experienced and successful producer manager can take the role of the diplomat or team-leader as described earlier. I have undertaken the team-leader role within my current PJS[3] production team, collaborating with an accomplished music programer and talented rhythm composer. The final element that PJS Productions lack towards my ideal pop songwriting and production team is a 'top-liner', which is the music industry term for the lyricist and vocal melody creator and collaborator. Most often, the top liner will also be an experienced vocalist.

Zagorski-Thomas (2014) states that the cultural domain's taste and function are the key factors behind forming music industry business models, such as PWL and P&E Music. My commentary on the forming and function of both 'brands' has shown that the complexities of aesthetics and authenticity play a significant role in the sustained creative and commercial success of these types of business models. Throughout the history of modern recording since the 1950s the balance of power has changed between economic capital and social, symbolic and cultural capital (Bourdieu, 1984). The awarding of MBEs to The Beatles in 1967 marked their recognition by the political establishment as pop cultural capital. Around the same time, The Beatles producer, Sir George Martin, negotiated royalties for record producers and laid down a foundation for future music producers by establishing his own independent recording studio (Air Studios) and production company. That created an economic capital shift for record producers and independent recording studio owners that still exists today, forcing the larger record companies to invest in the cultural capital that independent studios, producers and artists now owned. These shifts and the ever improving recording technologies paved the way for models such as PWL in the 1980s and P&E Music in the 1990s. Nevertheless, the cultural establishment has always been quick to criticize those models' successes and these types of artistic criticisms of successful pop music have existed for a long time. Adorno seemed to hold the elitist belief that nothing can be both popular and artistically valuable, stating 'the composition hears for the listener' (Adorno, 1941) suggesting that no listening effort is required for popular music. In Adorno's *On Popular Music* (1941) essay he describes a system that has remarkable similarities to the aim of the manufactured pop music of the 1990s:

> *Structural standardization aims at standard reactions.* Listening to popular music is [a manipulation into] a system of response mechanisms wholly antagonistic. In serious music, each musical element, even the simplest one, is 'itself'. For the complicated in popular music never functions as 'itself' but only as a disguise or embellishment behind which the scheme can always be perceived.
> (Adorno, 1941, 2002, p.442)

It is easy to perceive that although Adorno's description of popular music is from the 1940s, he may well have been talking about the manufactured pop and BoyBand music of the 1990s. The disguise and embellishments that producers use in pop music correlate to the 'plots and referencing' schemes that I have described in the compositional stage (Chapter 4) and the production stage (Chapter 5) of manufactured pop music. These referencing techniques then become apparent in the structure of the creative part of my 'Service Model of (Pop Music) Creativity and Commerce'.

8.2 A 'SERVICE' MODEL FOR (POP MUSIC) CREATIVITY AND COMMERCE

The central issue of this model is to encourage collectivist rather than individualistic thinking. With empirical analysis I have ethnographically reflected on the creative production workflow system at PWL Studios in the 1980s and the way in which P&E worked with manager and entrepreneur, Tom Watkins. The result is the service model, which combines my pop songwriting and production frameworks outlined in Chapters 4 and 5 and further analyzed in section 2 of the Appendix. I would agree with Sawyer's (2003) statement that 'evaluation must occur at the ideation stage' during group creativity sessions and this reflects pop music production and group collaboration in pop songwriting. Pop songwriting teams may discard five ideas towards a song within an hour but keep one that everyone agrees on and then move on to the next section of the song, which is a clear example of continuous evaluation by the team during a creative process. Therefore, such collaborative groups need to be comprized of like-minded creative individuals who possess different but complementary skill bases to contribute to the project. Keith Sawyer describes creative interaction that is based on Csikszentmihalyi's (1997) systems model for creativity and fits closely into my 'service' model for (pop music) creativity and commerce:

> The domain is constituted by the set of ostensible products created by members of the field, and selected by the field to enter the domain through a collective, social decision-making process. The creative process involves the generation of a novel product by the individual; the evaluation, or 'filtering', of the product by the field; and the retention of selected products by adding them to the domain. Thus the creative process involves a continual cycle of person – field – domain – person, which is mediated by ostensible, more or less permanent creative products.
>
> (Sawyer, 2003, p.123)

Creative interaction in the pop and BoyBand domains do not quite work as Sawyer has described but can easily be adapted or interpreted by suggesting that the 'team-leader' takes responsibility for all the decision making, evaluation and filtering. I am suggesting that the team-leader is the person at the beginning and the end of Sawyer's suggested creative cycle (see Figure 8.1). Whilst Sawyer (2003) separates group creativity and product creativity as two different entities and forms of creativity, in manufactured pop production they are viewed as one. I have described this in the process of P&E songwriting towards our 1990s pop projects earlier. Entrepreneurs and creative individuals that I have described throughout the book, such as

Tom Watkins and Pete Waterman, act as gatekeepers to the field and the domain in these types of creative songwriting situations. They are what I have called 'team-leaders':

> I am telling you even now I still have the ability, when I listen to a record to know instantaneously whether it is a hit record or whether it's fucking garbage; you just know.
>
> (Watkins, 2014, personal interview)

Could we call the 'knowing' that Watkins describes as a visceral truth, a certainty deep in one's bones that a hit record is being created in the recording studio? Vitally, the team-leaders take on this role of responsibility at the conceptualizing of the pop song or production and also at the end mixing stage, generally leaving the other skilled individuals creating music in the recording studio field, time and space to express themselves. Certainly that was the style of team-leadership I observed of Pete Waterman's role within the SAW songwriting and production team at PWL Studios in the 1980s. Whereas my experience with Tom Watkins guiding P&E Music in the 1990s involved Watkins spending countless studio hours with Ian Curnow and I compared with Waterman at PWL. Each team-leader will have a different style that will suit their personality and will indicate the level of control over a project that they wish to assert. During my interview with Tom Watkins he admitted asserting 'complete control' of every aspect of the East 17 and Deuce records from the creative production to the marketing of the product to the commercial domain. The role of the team-leader is to continually have their eyes (and ears) on the creative aspects and the commercial potential of a project. In this music genre, the commerce will take priority over the artistic though, because the team-leader knows that the songwriting and production team will not exist without the financial rewards of the group collaboration. Watkins states his view on when the model may fall apart:

> Once a manufactured band achieves vast success, they often forget the other players who helped to make them big. They view themselves as the sole architects of their fame and glory. And it usually ends in tears.
>
> (Watkins, 2016, p.285)

Ian Curnow and I experienced this kind of attitude progressively growing each time that East 17 visited our P&E Studio within The Strongroom Studios. The band members discussed their previous successful records in a language that described them as the creators of their own productions and mixes, even though the majority of the band members would only be present for the vocal sessions. It seems inevitable that this service model will only last for a certain amount of time for one manufactured pop or BoyBand but the strength of the model is that the songwriting and production should have the ability to provide success for multiple future acts as long as the team-leader can keep everybody happy with a continual workflow. I am suggesting therefore that the pop music songwriting and production teams directly manage the intersection between creativity and commerce, which differs slightly from Richard Burgess's view that 'music producers manage the intersection of technology, art, people and commerce' (Burgess, 2014, p.178). The technicians in the team are managing the technology, everyone is managing the art throughout their workflow and the team-leader is managing the people, the creativity and the commerce. I am also therefore suggesting that the whole manufactured pop and BoyBand process would be difficult for one person to achieve alone, especially long-term. On interaction and creativity, Sawyer (2003, p.125) states that 'the study of group creativity can help us understand how social and interactional processes contribute to product creativity'. The following section looks into group creativity using my pop songwriting and production service model.

The flow chart in Figure 8.1 highlights the ideal agents, field and domain for a service model for (pop music) creativity and commerce. The methodologies of the songwriting, production and mixing stages have been discussed in Chapters 4, 5 and 6, as well as further details in the

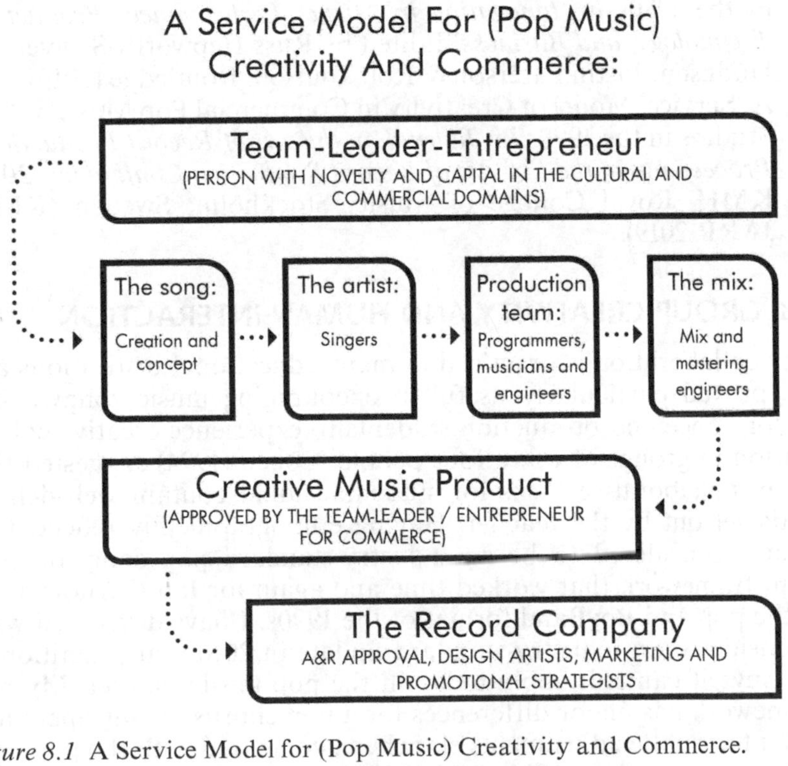

A Service Model For (Pop Music) Creativity And Commerce:

Team-Leader-Entrepreneur
(PERSON WITH NOVELTY AND CAPITAL IN THE CULTURAL AND COMMERCIAL DOMAINS)

The song: Creation and concept

The artist: Singers

Production team: Programmers, musicians and engineers

The mix: Mix and mastering engineers

Creative Music Product
(APPROVED BY THE TEAM-LEADER / ENTREPRENEUR FOR COMMERCE)

The Record Company
A&R APPROVAL, DESIGN ARTISTS, MARKETING AND PROMOTIONAL STRATEGISTS

Figure 8.1 A Service Model for (Pop Music) Creativity and Commerce. *Illustrated by Vince Canning, 2018. Copyright Phil Harding 2018.*

Appendix. The most relevant points of these methodologies are priorities such as the right choice of team-leader. Often the team-leader will start the process and assemble the appropriately skilled agents around them as I have described with Pete Waterman and the SAW team of the 1980s. With P&E Music and Tom Watkins, it was the other way around: Ian Curnow and I already existed as a creative pop songwriting and production team with good cultural capital credentials and enough economic capital from our 1980s PWL work to afford our own, well-equipped, studio space at The Strongroom Studios. What P&E lacked was a team-leader who had all three of the Bourdieu (1984) sociological success ingredients: Cultural, symbolic and economic capital. Importantly, P&E also required a team-leader with novelty in the cultural domain (Csikszentmihalyi, 1997) and Tom Watkins provided all of those ingredients with abundance. Commercial success (not necessarily chart success) could be achieved by 'service agents' from 2019 onwards who choose to reference the 'Phil Harding Pop Song Arrangement and Production Methodologies' outlined in Chapters 4 and 5.

Two further examples of the 'Service Model in Action' have been co-written by Dr Paul Thompson and myself and published as conference paper proceedings chapters:

1. A 'Service' Model of Commercial Pop Music Production at PWL in the 1980s in *'Innovation in Music: Performance, Production, Technology, and Business'* Edited by Russ Hepworth-Sawyer, Jay Hodgson, Justin Paterson & Rob Toulson. Routledge (2019).
2. A 'Service' Model of Creativity in Commercial Pop Music at P&E Studios in the 1990s in *'Journal on the Art of Record Production – Proceedings of the 12th Art of Record Production Conference, 2017'.* KMH, Royal College of Music, Stockholm: Sweden (KMH/JARP, 2019).

8.3 GROUP CREATIVITY AND HUMAN INTERACTION

Peer collaboration is practiced in many educational institutions and has proved particularly useful in encouraging music songwriting, performance and production students to experience creative collaboration in groups of up to four people. Cohen (1994) suggested that group collaborative tasks for students should contain well-defined goals set out by the teacher. Having ethnographically reflected on what I considered to be an industry-standard pop song arrangement framework that worked time and again for Ian Curnow and I in the pop and BoyBand format of the 1990s, I have developed what I considered an industry-standard update in 2016 that practitioners like myself can still apply today in the pop music market. My new framework has major differences for a pop chorus arrangement and just a few subtle adjustments elsewhere, compared with the pop song arrangements of the 1980s and 1990s.[4] I have had the good fortune to test the new framework in a case study during the period of this research at my aforementioned songwriting and production mas-terclasses in Oslo during February 2015 and February 2016 at The Westerdals School of Arts, Communication and Technology. Course leader Jan-Tore Diesen presented me with sixteen students for 48 hours with the intention of dividing the students into four groups of creatively compatible songwriting and production teams that would, with my instruction, write and produce a new song for one of my current production artists. Having introduced myself and set out the plans and goals for the next 48 hours, the precise instructions for the songwriting task can be found in Chapter 4.

Each student team consisted of at least two musicians, generally a guitarist and a keyboard player, plus a lyricist and a creative techni-cian/confident music programer. The teams were left to choose their own 'team-leader' or session producer that would take responsibility for the team's retention of the framework guidelines throughout the sessions. Each team was given a separate production suite to work in for both days. For the second day a professional female session vocalist was hired and I guided the teams through the production process of interacting with the singer, under time-pressure, to record lead and backing vocals (as described in Chapter 5) appropriately for industry presentation. What emerged was surprising; one would think that the song reference guidelines (tempo, key, references and song arrangement) were so strictly defined that all four songs would sound the same or at least similar. The opposite of this was the case. This demonstrates that when human interactivity is introduced to the creative pop songwriting process, the results will never be the same. 'Group Creativity' as Keith Sawyer (2003) calls it:

> These effective collaborating groups manifest emergence – the outcome cannot be predicted, the whole is greater than the sum of the parts.
>
> (Sawyer, 2003, p.185)

Resolving conflicts, making decisions and solving problems are the type of situations that students will face in their future careers in the creative industries and employers are increasingly interested in ev-idence of these skill bases and collaborative experiences. Jan-Torre and I would continually visit the four different groups throughout the first day of this task as they can easily 'fall from a *flow* state into anx-iety' (Sawyer, 2003). The creative *flow* process of *human interaction* within the group is vital to achieve the required results and the feed-back shown in Chapter 4 from my music industry client, Carl Cox of Prolific Music UK, has been useful for the students to internalize for their own self-analysis. The ultimate accolade of this framework and process is that Prolific Music has commissioned a master recording of one of the four songs presented. This has given the students in that team the opportunity to experience negotiations with a UK music industry publisher and production company.

8.4 POP AND BOYBAND PRODUCTION: THE SONIC PICTURE

There are many different ways to analyze and discuss production techniques and working practices. During my many years as an industry practitioner I have developed a technique that has been touched on by others, including the record production academic Simon Zagorski-Thomas. In his book *The Musicology of Record Production* (Zagorski-Thomas, 2014), Zagorski-Thomas uses the term 'sonic cartoons' whilst discussing music and metaphor, comparing the cartoonists' visual formation of one of the characters and how it affects our judgment of that character to how audio engineers and producers might treat a sound performance or 'sonic image' that could be a single musical part within a production.

Producing a pop record has for me, for some time, been comparable to painting a picture. I see pop production as constructing a large, complex, layered canvas, where it is not uncommon for there to be well over a hundred audio tracks to deal with in a mix. UK producer and engineer Haydn Bendall (2015) stated to students at an industry lecture 'a mix is never finished – just abandoned (hopefully at the right point and time)'. The brush strokes applied by the musicians and the producer are never really finished or complete until the producer or team-leader says so and decides that the painting has reached the original vision that was imagined by the producer (and hopefully the artist) at the time of the first instrument recorded for the piece. This highlights a good comparison point of how I view a whole stereo mix or production. Usually, in the planning or early stages of a recording or production of a track I will have a 'sonic picture' in my mind of how the finished stereo mix will sound. The process of the initial recording, whether it be live performances by musicians that I am recording or programed parts into a music sequencer, becomes a sonic parallel to painting by numbers. I strongly believe in this process for successful pop music aimed at commercial and national radio play in the UK and Europe; this worked especially well in the pop and BoyBand genre of the 1990s and is still relevant today. This may sound a strange comparison point, but the reality is less 'painting by numbers' and more towards a contemporary modern artist who may have a sketch in their mind when embarking on a new painting but will allow changes and influences as the work progresses. Some might say it is as mechanical as painting by numbers but I always keep an open mind to accommodate suggestions, artistic compromises and commercial requests (usually from the record company or manager). If no one else on the project makes any suggestions or requests, the producer should have the ability to continue forwards through the recording and overdubbing stages of formulating a record that is artistically cohesive and commercially suitable. I was fortunate to train under an excellent mentoring producer by the name of Gus Dudgeon[5] who worked by this method. What I observed from Gus's production methodology was an undying certainty that his vision for the record would work, both for the artist and for the record company and nothing would step in his way to deviate from that path. Such is the balancing act that music producers live with every day.

Therefore, when starting a production I will allow interpretations of the planned ideas from the musicians and programmers I am working with if it 'feels' right for the track, so it is an ever-changing sonic landscape but there is an underlying plan which will always be the foundation of the track. The foundation, things such as tempo, key, song arrangement, timbre, basic top-line melody and 'plot' references will generally always remain.

> Timbre is a function of the nature of the object making the sound as well as the nature of the type of activity.
>
> (Zagorski-Thomas, 2014, p.65)

Whilst one can find and interpret many meanings and uses of the word 'timbre' in music, such as resonance, character, quality of sound, tone, reverberation or pitch, none seem to talk about the *feeling* that a producer, creative technician or musician has when they know something sounds the way they wanted to hear it. All of the individuals within a team are affecting the system by way of their creative opinions on the musicality and the technology in place at the time of recording. Meanwhile, during the final stereo mix session process one can talk about timbre and the way a particular overdub is treated by staging in the spatial sense (Moylan, 2015 or the various reverbs and other processing effects that can change the perception of that individual part. Again, this is driven by the *feeling* of what the mix engineer wants to hear – or is perhaps being asked for by the producer if there is one present at the session.

The opposite or alternative production technique to my 'painting by numbers' approach is often called the 'organic' approach, whereby the producer sets about the project with no specific end plan and will start with the initial song arrangement, as presented by the artist or songwriter, and allow the production and arrangement to develop in the studio, listening to the views and performances of the musicians involved and allowing everyone to make 'group' decisions throughout the journey. The theory is that this will grow naturally as the session progresses and the end result will be an artistic and diplomatic result, often taking in the views of technicians as well as the musicians, singers and artists. I file this approach away in a drawer for reference as I view this to be a potential recipe for commercial disaster.

I would recommend that anyone entering the manufactured pop or BoyBand production arena consider the 'Service Model for (Pop Music) Creativity and Commerce' described in section 8.2 and shown in Figure 8.1. This sets out a working model for production teams when combined with the techniques described in Chapter 5. My commentary on East 17's 'Stay Another Day' single in Chapter 7 is a good example of the service model in practice as a team is working at *peak flow* and although I have described the problems of capturing Brian Harvey's lead vocal performance to the required level, the team consisted of enough skilled agents in the field to assist in achieving the desired result. The commentary on the problems encountered with the East 17 production 'Someone To Love' in Chapter 7 highlights the extreme levels of 'service' that a production team may need to employ to achieve a successful completion of the service model. Watkins (2016) describes this domain as 'the giddy carousel of pop' and he was a great example of my service model's team-leader throughout the four years that P&E worked with him (1992–1995). Sometimes a successful production team can reach such an extreme version of *peak flow* where it would appear that they can do nothing wrong and could even pluck an unknown singer from obscurity and deliver a #1 single. I witnessed that phenomenon at PWL in the 1980s when Pete Waterman signed the singer, Sonia, after she approached him with a demo tape outside his Liverpool radio station. In 1989, within 2 months of that approach, Sonia and Waterman achieved a UK #1 single with the SAW composition 'You'll Never Stop Me From Loving You'. I have described throughout this book that I view the requirement for a creative team's *peak flow* to be dependent on all the links in the creative and commercial chains to be working at their highest levels and another example of this is the #1 record that P&E produced for Boyzone with the cover of the Bee Gees song 'Words' in 1996. My full commentary and reflections on the Boyzone 'Words' production are detailed in Chapter 7, showing that Ian Curnow and I were back to *peak flow* (after parting company with Tom Watkins in 1995) and all of the promotional, commercial and social elements surrounding Boyzone were perfectly aligned for the record's success. Tony Mortimer of East 17 likened my 'all the links in the chain' analogy to his view that successful pop records have to start with a great song as the cog of a wheel. All the spokes that connect to that wheel must then be correctly aligned to successfully complete it, making it possible to create a hit record.

8.5 CREATIVE PERSONAL *FLOW* IN POP AND BOYBAND MIXING

The poet, Mark Strand, interviewed by Mihaly Csikszentmihalyi, describes forgetting the self and sense of timelessness in creative *flow* when things are going well:

> Well, you're right in the work, you lose your sense of time, you're completely enraptured, and you're completely caught up in what you're doing. The idea is to be so saturated with it that there's no future or past, just an extended present.
>
> (Mark Strand in Csikszentmihalyi, 1997, p.121)

We could call this entering a different reality, where existence is temporarily suspended. This is called a truism in the study of creativity. *Flow* like this can happen in many other creative arts, sports, mathematics and business. On how it feels to be in *flow*, Csikszentmihalyi's (1996) research suggests that intrinsic motivation; whatever produces *flow* becomes its own reward and timelessness, being thoroughly focused on the present, where hours seem to pass by in minutes. Csikszentmihalyi (1992, p.71) states 'self-consciousness disappears, and the sense of time becomes distorted'. That is a good description of the *flow* in my mixing process and is something that I have recognized whilst practicing my 12-step mix system from the 1990s onwards. During my mixing work from the 1980s onwards I have always used a sonic landscape in relation to an artist's picture landscape. To take that further in the mixing stage one needs to visualize the picture as 3D so that one can analyze the staged layers in a deliberate attempt to initially separate the instruments for the listener and yet also help those instruments to blend together as one. The aim is to achieve a mix that sounds and feels to the listener as though all the musicians (live or programed) and singers are in one space or on stage together. My domain for this mixing has generally been in the pop and dance music genres, largely dominated by manufactured pop and BoyBands of the 1990s. I was supported in the field by well-equipped recording studio control rooms[6], assistant and maintenance engineers, as well as my production associates (SAW) and my 1990s production partner, Ian Curnow. Csikszentmihalyi (1996) describes some of the requirements to achieve cultural success:

> To move from personal to cultural creativity one needs talent, training, and an enormous dose of good luck. Without access to a domain, and without support of a field, a person has no chance of recognition.
>
> (Csikszentmihalyi, 1996, p.344)

I did not enter the music industry in 1973 to become a specialist pop music mix engineer; I had hoped to work in the rock music genre, which I was an admirer of at that time. The specialist music genre that creative technicians find themselves in will generally be determined by their domain and field situations, often more by luck than necessarily choosing, as Csikszentmihalyi suggested. I would not necessarily say that I possess a huge amount of creative talent but I have had a lot of training (mainly by observing others in the field) and an enormous amount of luck in terms of initially securing my first job at Marquee Studios in 1973[7]. I then followed my 'gut instincts' on moving from Marquee Studios to PWL Studios with SAW in 1984 and that same instinct again on my PWL departure in 1992 to The Strongroom Studios with Ian Curnow. Both of those career moves were achieved due the music industry recognition of my cultural creativity successes at the time. During Csikszentmihalyi's interview studies for his book, *Beyond Boredom and Anxiety: Experiencing Flow in Work and Play* (Csikszentmihalyi, 1975), he discovered that one could not get to the automatic, spontaneous creative process or create anything successfully without ten years of training. With the technical knowledge and immersion in a particular field to develop the technique, today we may shorten that time from what I have witnessed through running songwriting workshops with young teenage songwriters with an average age of fourteen at Dutch Van Spall's Strawhouse Studio in Rugby, UK. Those songwriters displayed evidence of mature songwriting techniques and abilities beyond their years, showing the potential to attain successful careers within two to five years as opposed to Csikszentmihalyi's suggested ten years.

The music mixing process has never been underestimated by the recording audio industry. The reward of achieving a personal creative *flow* has largely come about during my time spent in the mixing stage of the field process. I would generally be alone and isolated for the majority of a typical mixing day, and as Csikszentmihalyi states 'when we live creatively, boredom is banished' (1996, p.344). Reflecting on this, I often say to students today, 'find your passion, make it your profession, and you will never have to do a hard day's work in your life'. This is a fine example of how I have viewed my life in the recorded music industry. I have never been bored and, especially in the mixing process, I have developed a disciplined routine for controlling the *flow* of my creative energy. That process (the 12-step mix program) is described in detail in Chapter 6 and it begins with placing the lead vocal at the front of my 3D sonic landscape. What appears to be a mechanical series of jobs that follow can only be achieved by keeping the perceived sonic landscape picture in mind at all times and managing a sensible personal time schedule so as not to lose the mental energy required to arrive at the goal. With all the elements in place, the personal creative *flow* in pop music mixing can be attained. Many times I have reached the optimal experience of Csikszentmihalyi's *flow* and for my current mixing process it has become an 'automatic, effortless, yet highly focused state of consciousness' (Csikszentmihalyi, 1996, p.110).

To provide a perspective of the economic capital that the music and audio industries apply to the mixing process, it is worth comparing the fee structures of freelance recording engineers and assistants in the UK music industry, which incidentally have not changed since the 1980s, to understand how industry values the specialist mix engineer:

1. Freelance assistant engineers, London, 2019: **£50–100 per day.**
2. Freelance recording engineers, London, 2019: **£100–250 per day.**
3. Freelance specialist recording engineers (full orchestral recording for instance) London, 2019: **£250–500 per day.**
4. Freelance mix engineers, London, 2019: **£500–1,000 per day.**

Whilst music producers will generally achieve higher figures than these examples per project, plus a royalty on sales, the in-demand mix engineer can also command a small share of the royalties with the appropriate negotiations. The music mixing schematics that I have described have inspired my reflections on the fact that I can generally recognize my mixes in various situations, such as radio play and social media play. It would be correct to say that Ian Curnow and I were synonymous with our manufactured pop and BoyBand sound that we co-created in the 1990s, the evidence of this is supported by my descriptions of the way in which Boyzone's manager, Louis Walsh, approached P&E to produce the same style of dramatic pop orchestral sound from our East 17 productions for the Boyzone single 'Words'[8]. This leads to a realization that I have used sonic characteristics that combine to make a 'Phil Harding mix' signature sound (Zagorski-Thomas, 2014). Described by Csikszentmihalyi (1990) and citied by Sawyer (2003) is personal 'peak experience':

> Mihaly Csikszentmihalyi described, 'flow' as a peak experience, a particular state of heightened consciousness. Extremely creative people are at their peak when they experience 'a unified flowing from one moment to the next, in which we feel in control of our actions, and in which there is little distinction between self and environment; between stimulus and response; or between past, present and future'.
>
> People are more likely to get into *flow* when their environment has four implicit characteristics:
>
> 1. They are doing something where their skills match the challenges facing them.
> 2. *Flow* occurs when the goal is clear.
> 3. When there's constant and immediate feedback about how close you are to achieving that goal.
> 4. *Flow* occurs when you're free to concentrate fully on the task.
>
> (Sawyer, 2003, pp.39–42)

The optimum evidence of not only reaching 'peak experience' but also achieving your goals as a mix engineer and music producer is that the product is commercially successful in the cultural domain it has been designed for. Chart positions, sales, radio play and acceptance by specialized critics are all signposts that the creative energy expounded to reach that peak experience was worthwhile. My Chapter 7 descriptions are good examples of 'peak experience' *flow* in my mixing and the 12-step mixing program in Chapter 6 sets out a system to achieve this. The fact that both of these songs ('House of Love' and 'Stay Another Day' by East 17) achieved good chart performances as well as critical acclaim at the time, further highlight the phenomenon of Csikszentmihalyi's *flow*. I remember a feeling of transcendence and a state of joy at the end of the week's work before Christmas 1994, knowing that 'Stay Another Day' by East 17 was #1 in the UK singles charts for the holiday season and nobody was going to remove that euphoric 'moment in time' reflective memory from me. This was also a good example of creativity and commerce coming together as one, knowing that my involvement in the project was total. It is difficult to summarize a music producer's role within a successful music project but Burgess (2014) succinctly states:

> In all its permutations, music production is – at its essence – an art form, whose goal is to produce a unique sonic artifact that captures the vision of its creators, the imaginations of its audience and that will service the needs of its stakeholders.
>
> (Burgess, 2014, p.179)

In terms of creativity and commerce, the Burgess scenario is an ideal, but in the world of manufactured pop and BoyBand music I have highlighted the case for one to interpret the scenario of a creative team, guided by the team-leader, to warrant the title of 'the creators', or artists, together with the featured artists to take the equal plaudits from the audience and the stakeholders.

8.6 THE PHENOMENON OF OPTIMAL EXPERIENCE

So how can we achieve the ideal optimal experience as songwriters, creative technicians and pop music producers today? Csikszentmihalyi (1992, p.83) suggests that it is four times more likely that we will achieve the *flow* experience at work as opposed to watching television at home. Consumer technology has changed significantly since 1992 and the interaction that players of computer games experience would allow justification for them to claim an optimal *flow* experience, especially those who interact with others during online game-playing. Csikszentmihalyi (1992, p.83) also states that 'it is not easy to transform ordinary experience into *flow*'. Zagorski-Thomas (2014) seems to favor the actor-network theory (ANT) over Csikszentmihalyi's systems model for creativity as he views ANT as a better tool 'to explain both the network through which recording technology is produced and disseminated and the network in which it is used to make music' (Zagorski-Thomas, 2014, p.92). The ANT system can therefore be used to describe the participants, their environment and their relationship to each other within a music production. I have naturally and unknowingly adopted this system throughout the book and the associated commentaries to describe the pop production process. I have provided in-depth descriptions of the process, the social aspects of the production team, musicians, singers and the recording studio environment (in its many modern forms). The roles I have described within the PWL and P&E teams from the 1980s and 1990s allow a clear understanding of my current (2019) PJS productions network. The roles are clearly defined and match my 'service model for (pop music) creativity and commerce':

1. Team-leader: Phil Harding.
2. Music programer: Julian Wiggins.
3. Rhythm programer: Simon Dalton.
4. Musicians: Greg Fitch (guitar)/Davide Trottorius (bass).
5. Singers: Melissa Wiggins and Liz Richardson.

There will always be an element of configuration between these people – asking, showing, persuading, telling and sharing. Zagorski-Thomas (2014, p.152) also discusses 'notions like power, roles, persuasion and trust' around a collective activity in ANT and again this theoretical approach is captured well in my discussions around a service model for (pop music) creativity and commerce, see Figure 8.1. Throughout the 1980s and 1990s my descriptions of the environments for these activities have centered on creative hubs, such as PWL Studios and The Strongroom Studios. In 2019, creative hubs, such as Tileyard Studios in London and Confetti in Nottingham, have arisen in many cities in the UK, attracting a variety of creative industries that complement music recording. The speed of internet technology and compatibility of music technology with music sequencers, such as Logic Pro, has allowed teams, such as PJS Productions, to achieve professional results without requiring a physical collaboration space, provided each participant has a home studio with a fast internet connection and file-sharing platforms such as Dropbox. The cultural domain of this PJS example is the set of rules, conventions and knowledge agreed within the boundaries of the team, often forgotten or misunderstood in daily practice, but they are always there in the background or the underlying structure and foundation of the team. In 2019 there is now a fine balance between economics and art in music production. Record company budgets are at an all-time low because the label executives know that most pop production teams are structured in a similar way to PJS, where the necessity to hire commercial studios is vastly reduced because the majority of producers and musicians now own a home studio. The complete recording budget is therefore subcontracted to the production team who in turn will hire session musicians as they are required. In theory, the art should not suffer because of the economics, due to there being few time-constraints (within reason) for the musicians and producers to perform within this framework. Another alternative systems model that takes a constructionist approach to looking at technology is the social construction of technology (SCOT) method. This looks at the mindset of the social participants, as I have done throughout this study, particularly where my interview respondents and I have described the team-leaders, such as Pete Waterman and Tom Watkins.

We may not agree on how something should be done but we do agree on the context for framing the question and discussion on how to work together as a team to achieve our creative goal, using technology and performance. The cultural domain within the SCOT model sets the boundaries of how we use our technology and interact with each other, the action script, to achieve a creative and commercial result.

This empirical study captures the 'lived experience' (Husserl, 1927) of the interview respondents and my own reflections throughout the 1990s pop and BoyBand phenomenon. The interpretations of my respondents' extracts have assisted the analytical commentary throughout, allowing my own reflections to flourish and develop the two theories presented. The research has provided a unique insight into creativity and commerce within the context of manufactured pop and BoyBand music. My service model for (pop music) creativity and commerce offers a clear outline of the process, personalities and elements required for current and future pop production teams. The observations of placing the service model into practice with students in Oslo has led to the theory that human interaction will affect group creativity in unpredictable ways and this builds on the work of Sawyer (2003, 2007). The results of the 2015 Oslo sessions featured in Chapter 4 were completely unexpected and can be heard here – https://soundcloud.com/user-606698363 – to demonstrate the differences between them. The industry feedback from Carl Cox of Prolific Music has provided an independent analysis of the creative results from a commerce-driven perspective. I have explored the idea that creativity is a product of the interaction between three main elements of domain, field and agent (Csikszentmihalyi, 1996; McIntyre, 2007; Kerrigan, 2013) and have found that in the context of manufactured pop and BoyBand music it does not allow one to prioritize the 'team-leader' in the way that I have highlighted in my service model for (pop music) creativity and commerce. Certainly, the Csikszentmihalyi model is at play in manufactured pop and BoyBand music, and certainly, Kerrigan's idea that creativity is the central product of Csikszentmihalyi's system model but neither of these systems encapsulate the intensity and impact at play when the team-leader controls the domain, field and agents in the way that I have described with Tom Watkins and Pete Waterman. A principal aim of this book was to highlight the vital role of the 'team-leader' in the collaborative projects that I have described throughout Chapter 7 and the distinct forms of capital (Bourdieu, 1984) that person will possess. I have been in a unique position to collaborate with and observe these team-leaders working at close quarters in the recording studios during the 1980s and 1990s. By the late 1990s I was experimenting with adopting the team-leader role myself and have now taken on that role for my more recent productions with PJS.

Finally, this book has illustrated how the technology in the early 1990s affected the creative results on pop studio projects, such as the 'House of Love' single by East 17 (see Chapter 7) and how computer and MIDI technology in the 1990s transformed the working methodologies for music producers in this field and domain. The commentaries and descriptions of songwriting arrangements, production techniques and my 'top down, 12-step mix program' from the 1990s through to 2019 can provide further insight into how teams of future pop music creators and collaborators might adopt a service model system for their work and reach the phenomenon of optimum experience.

NOTES

1. Matt [Aitken] and I, throughout all of our success, *were* the band. We performed on all our records. We were songwriters, we were a band, we were musicians, we were the artists really and they [the credited artists, such as Kylie Minogue] were the guest singers. Nobody has really given us credit for that (Stock in Egan, 2004, p.283).
2. See *PWL From The Factory Floor* (Harding, 2010).
3. PJS Productions consist of Phil Harding, music programmer, Julian Wiggins, and rhythm composer, Simon Dalton.
4. The outline of that new arrangement is described in Chapter 4 under the heading 'Phil Harding Song Arrangement Methodology Example 2019'.
5. Producer of the majority of Elton John's 1970s repertoire, including the 1973 classic album 'Goodbye Yellow Brick Road'.
6. See the Appendix for studio equipment details.
7. The first chapter of my book, *PWL From The Factory Floor* (Harding, 2010), describes these circumstances in detail.
8. See Chapter 7.

Appendix: Technology for Pop Music Production

A.1 INTRODUCTION – TRANSFORMATION AND PROCESS

Richard James Burgess (2014) states that 'transformative technologies', such as digital audio workstations (DAWs), completely changed the working practices of creative studio technicians and producers, such as Ian Curnow and myself, in the 1990s. I commentate on that throughout this appendix and how the successful workflow of some musical creatives, such as Stock and Aitken, can cause a reluctance to change as these transforming technologies arrive. Time will usually cause an eventual shift to new technologies for pop production teams though because artists and commercial clients will be concerned about working with out-of-date producers using incompatible technology to their rivals. Burgess (2014) also states that by the early 1990s the consumer market was also affected by 'transformative technologies' with the arrival of the CD, which quickly surpassed the sales figures for vinyl. Fast on the heels of the consumer CD we had the arrival of what many technologists would call a 'disruptive technology' in the shape of the compressed, low-quality MP3 download music file. The problem for professional technologists was that this format, developed by the Fraunhofer Institute for Digital Media Technology in Germany, took us backwards in terms of a commercial format delivering to the consumer a high-quality representation of the music productions we had spent many hours of studio time perfecting.

Simon Zagorski-Thomas (2014) states that the common 'cut and paste' methods of DAWs have encouraged composers to work in a modular fashion since their inception in the late 1980s into the 1990s, causing producers to adapt their methodologies and record session musicians in the same way. This means to only record one chorus and to cut and paste that take throughout the arrangement. This has become completely normal practice for the pop and BoyBand music genre I am describing throughout the book. The lists of the P&E Music 1990s studio equipment at the end of this appendix were, in my view, the minimum requirements for a competitive pop production team of the day. Today, the equivalent technology can be found inside computers as virtual plugin copies of the hardware I will describe in this appendix.

Many pop producers were working on tight budgets in 2016 and would often incorporate offline collaborations with session musicians working from files accessed using online methods, such as Dropbox and WeTransfer. This requires musicians that are technically capable of recording themselves to professional standards, to satisfy the producer. I still use Pro Tools software today, largely for mixing, but I have also recorded many acoustic and live rock bands over the last ten to fifteen years using this efficient software. It still seems to be the best music software platform for recording and mixing though many industry professionals prefer Logic Pro as a music software program for composition and pop/dance programed music. Today, I tend to use Logic Pro with my current music production team, PJS. We compose, program, record and produce everything in Logic Pro and then export the full production as audio wav files to Pro Tools for mixing[1]. This has also, for many, become common industry practice for music production, sound design, music for media, television and film.

Music Production Technology Developments Throughout the 1990s

In the 1980s, music technology gradually shifted from largely unreliable hardware platforms with limited back-up technology, such as floppy discs (the largest capacity was 1.9Mb on average), to the bizarre Atari ST computer, originally designed for home gaming. The two popular music sequencer developers of the time were Steinberg (Cubase software) and C-Lab (Logic software), both developed their MIDI sequencer platforms utilizing the flexibility of the Atari computer. There were other MIDI sequencers on the market in the late 1980s but in the UK and Europe, these two dominated in recording studios and programing rooms into and throughout the 1990s. A full description of the music technology used at the very successful PWL Studios throughout the 1980s is accessible from my book, *PWL From The Factory Floor* (Harding, 2010).

Music Production and Technology

Burgess (2014) summarizes the way that creative technicians and producers feel about the recording studio being interactive and by stating 'production is a composition in sound' (Burgess, 2014, p.43). In the hardware and computer-programed pop world of the 1980s onwards it became commonplace for songwriter/production teams to achieve all their songwriting and compositional duties in the recording studio. Prior to this it was more likely that pop songs would be written in less expensive 'demo' studios or at people's houses as opposed to the same studio where the full master production would also be recorded.

Mike Stock and Matt Aitken of SAW confirm this from their 1980s hit-making period:

> **Stock:** Good songs for us are kind of invisible from the record. We rely on the studio.
> **Aitken:** We only ever once sat down to write songs when we weren't actually making a record, and that was much later on. There's only one occasion where we actually wrote songs in a void where it wasn't recorded as we did it.
> (Stock and Aitken in Egan, 2004, p.289)

Stock and Aitken confirm that to them, the recording studio was part of their songwriting methodology, a tool, as many practitioners like to say and that to sit down to write songs outside of the recording studio environment would have felt alien to them in the 1980s and 1990s. Ian Curnow was the person arguing for change in the PWL building in the late 1980s, suggesting to the other producers and programers to move away from the Linn 9000 drum and MIDI sequencer hardware machine and onto the Cubase/Atari computer platform. From day one of Ian's arrival at PWL in 1986, where he was originally hired to run the Fairlight system in the sub-basement studio for SAW, we saw him combining his Steinberg Pro 24/Atari system effectively. Ian successfully managed to run the Fairlight and Steinberg systems 'in-sync' and this impressed everybody collaborating with him, especially Pete Waterman, who seemed less resistant to technological change than Stock and Aitken. It took a long time for Stock and Aitken to embrace this technology and they would stand by their Linn 9000 system, described by Burgess (2015) as a 'self-contained production box' for many more years. Aitken explained this in an interview with Sean Egan for the book *The Guys Who Wrote 'Em* (Egan, 2004):

> We were still using a relatively archaic system [Linn 9000] when everybody else was using Atari-based systems and we spent so long trying to get a system that was reliable and having [ironed] the bugs out of the system over two or three, four years, there was little point going to something that was going to crash on you.
> (Aitken in Egan, 2004, p.302)

Aitken was correct to indicate that the early versions of the Steinberg Cubase music sequencer system would regularly crash on the Atari computer. This was understandably disruptive to the creative workflow of songwriting and recording sessions at that time. Ian Curnow's view was different and here is his description of the Atari/Cubase system circa 1991:

> 2Mb of RAM, that was the biggest system available [the basic was 1Mb] and I think the hard disk [which was purchased separately]

was about 20Mb, or maybe 40Mb, I'm not sure. Amazing! Cubase was a revelation, and having been involved with Steinberg as a Pro 24 user, they asked me, along with several other users, exactly what we'd want from a 'clean sheet' sequencer, having used Pro 24. So when Cubase came out, it was like it was made for me, so many of my suggestions and ideas were there for me to use, which was fantastic.

(Curnow in Harding, 2010, p.555)

Ian Curnow was an early Cubase beta-tester for Steinberg and therefore input many suggestions to the development of the software. The majority of my BoyBand songwriting and production work throughout the 1990s was achieved at the P&E Music Studio[2]. In early 1992, Ian and I were already running the most powerful version of the Atari/Cubase MIDI sequencer available to us and we invested in a 24-track analogue 2" tape machine to record our audio as professionally and as competitively (and to be compatible to other studios and producers) as possible. Record companies would hire us for remixing projects, sending copies of the original 24-track multitrack tapes and vice versa on our productions, where we would need to copy and send those to the record companies. In 1992 multitrack analogue was still the format of the day even though Sony and Mitsubishi had developed digital multitrack tape machines in the 1980s. Ian and I had used these machines at PWL Studios for some years but in the early 1990s they were still very overpriced (initially £100,000 approximately for the Sony PCM 3324-S 24track digital tape machine).

In the summer of 1992, Steinberg became the first developers to combine audio tracks and a MIDI sequencer in their Cubase system. To achieve this, Steinberg moved on to the Apple Power Macintosh computer platform, as the Atari system was too underpowered. P&E were amongst the first recording professionals to use this new Cubase Audio system in its beta version at The Strongroom. During 1992, Ian and I invested in the new Apple Macintosh computer combined with an early Digidesign D/A (digital/analogue) interface. This system was synced to our Soundcraft Saturn 624 24-track analogue 2" tape machine using the Digidesign Universal Slave Driver SMPTE unit.[3]

A.2 RECORDING 1990S MANUFACTURED POP AND BOYBANDS

In Chapter 5 I set out the typical recording schedules that P&E Music adopted at our recording studio throughout the 1990s for this music genre. Recording technology from the late 1970s and early 1980s had allowed us to potentially program everything on a pop record including drums, bass, synthesized keyboards, samplers, and so on, all driven using MIDI from a master keyboard. That leaves only the artist's vocals to record using the type of system I have outlined. I repeat Mike Stock's good point of the true nature of this type of creative scenario:

> Matt [Aitken] and I throughout all of our success *were* the band, we performed on all our records. We were songwriters, we were a band, we were musicians – we were the artists, really and they [the credited artists such as Kylie Minogue] were the guest singers. Nobody's really given us credit for that.
>
> (Stock in Egan, 2004, p.283)

Stock's description of the producer/musician's working practices in pop music are accurate and it was the same for Ian Curnow and I during our 1990s hit-making period with the manufactured pop and BoyBands. My technical outlines here drive home the point that the artists credited for making the records contribute their vocal performances but no instrument performances in 90–99% of the cases in this music genre. This was the methodology for the majority of vocal-led pop groups throughout the 1980s and 1990s and was also the working practice for artists at Tamla Motown throughout the 1960s. The famous Funk Brothers backing band would provide all of the backing music for the Tamla Motown singers. The shift from session players to the production teams of the 1980s onwards was the result of the technology available in recording studios to production teams that consisted of musicians and technicians. SAW and the technicians around them were a good example of this.

Ian Curnow and I recorded vocals at P&E Studios onto the 24-track analogue tape machine initially and would then transfer the vocal tracks required into the Cubase Audio sequencer using the Digidesign interface once the artists had finished at the studio. The ever-evolving combination of Steinberg Cubase Audio software and Digidesign hardware (now Avid) allowed us to grow throughout the 1990s from 4 tracks of audio to 8 tracks, and eventually 24 tracks of audio through a combination of more powerful analogue to digital/digital to analogue interfaces and updated Apple Macintosh computers designed to handle that amount of audio data, which required more powerful external hard drives and evermore complex data back-up systems. This vocal transfer process would kick-start a lengthy period of analogue to digital transfers, followed by many hours of vocal editing in Cubase to choose and compile the best vocal takes. Ian Curnow explains why he thinks this type of technical information is important to our knowledge:

> I get people asking me on Facebook for my Yamaha DX7 Bass sounds – I don't have them anymore! We used to be called 'International Rescue' in the 1980s and 1990s, where someone had a radio mix that they couldn't get right for the label or the manager

and we would be hired to save it. The process now is 'all in the box' [the computer] and revisiting tracks to make adjustments is a lot easier. You can re-visit/re-vamp tracks yourself without necessarily sending it off to another remix or production team to 'rescue'. We also tend to look for team collaborations and co-producers now rather than just handing the whole track over to someone fresh for a re-vamp. Technology allows us now to achieve the same processes as the 1990s but it's a lot easier now, we're swapping files and not tapes etc.

(Curnow, 2014, personal interview)

This is useful information for people who are trying to understand how BoyBand records of the 1990s were produced and gain knowledge of the equipment used to generate the sound sources. I also still receive messages on Facebook about the complete chain of sounds for instrument processing, such as our kick and snare drums from the 1980s PWL period. The P&E Studio equipment list in this appendix also contains references to their commonly used attributes, such as the Studio Electronics SE1 rack MIDI unit being used mainly for just bass sounds. I have asked most of my interview respondents a question that I often hear in academic circles: 'What affect does technology have on recording studio creativity?' I have related my question to the specific pop and BoyBand genre; here is a selection of those views:

Tee Green: One of the big things that we've had for some time is the ability to record one section at a time rather than making singers perform the whole song top to bottom. It has also helped that we can, since the 1990s, re-time and re-tune vocals and how that affects the vocal performances during modern recording.

(Green, 2015, personal interview)

Tom Watkins: Something that Trevor Horn was doing which was that big sound, they talk about records now.... I never hear records now that sound as big as those we did [East 17], which Trevor was an exponent of. I think we gave them a good run for their money. Even he [Trevor Horn] got a little bit snobby about what level he was going to take it on [the competition between producers]. I feel that some of the stuff that you guys did contributed so heavily to that whole make-up and fullness of sound [in the 1990s], which I think has been lost. Because recording budgets have now been capped.

(Watkins, 2014, personal interview)

Some interesting views on how technology has changed our working methods and perceptions of sound, especially in the way Tom Watkins talks fondly of his analogue recording experiences with P&E in the 1990s. The way that technology has developed from the late-1990s onwards, with the introduction of vocal-tuning processing software, such as Autotune and Melodyne, is the kind of assistance to vocal recording sessions that Tee Green refers to. All this technology enhanced the creative process for the pop and BoyBand recordings throughout the 1990s. The focus for this music genre is always on the vocals.

Recording Studio Etiquette

There are a number of unwritten rules for studio etiquette; the kind of behavior that is ideal or most suited to recording studios of any kind of size or any kind of situation. These rules have been touched upon by various authors commentating on recording studio methodologies (see Burgess, 1997; Massey, 2000, 2009; Visconti, 2007; Harding, 2010). The harshest of these would be for studio assistants, who are still needed today, to 'only speak when you're spoken to'. It is surprising how many people can make this error. Often it may be visitors dropping into the studio to say hello and that visitor may be unaware of the potential conflicts and diplomatic scenarios within the team that have been working on a project for days, weeks or even months. Richard James Burgess has been known to say that it may be wise for a music producer to have meditated before entering the melting pot that can be a recording space, overcrowded with artistic and creative egos. The two occasions when a production team such as P&E would have to prepare for these visits, and one has to prepare to create the right atmosphere for creativity, would be the vocal recording days and sometimes the mixing days.

Ian Curnow and I had a full JBL 5.1 sound system installed into the P&E Music Studio lounge area, to compliment the Phillips widescreen television and the laserdisc film playback system. It was the early days of this type of television system so it became a novelty for colleagues to visit us from within The Strongroom Studio complex as well. The main reason for installing this system was to create a contemporary technological environment for artists to visit that was not just a workspace.

A.3 ORCHESTRAL STUDIO METHODOLOGY FOR POP MUSIC OVERDUBS

Jamie Shaw Project (Decca Records): 1999 and 2000

This was a contemporary pop project featuring the singer, Jamie Shaw, who was aged fourteen at the time having been discovered on Michael Barrymore's Sunday night television variety show. Jamie had a powerful pop ballad voice, which became the reference for the direction of the album, medium to slow pop songs, some covers, some hymns but nothing classical such as the material heard at that time from Charlotte Church. Jamie Shaw and had been signed by Decca Records for all these strengths. Initially Ian and I had three tracks to orchestrate on the project in 1999 and at the record company's request this was going to require strings, woodwind and brass in one three-hour session. The UK Musicians Union only allows overdubs onto a maximum of three songs in a three-hour session and during that three-hour session there is a compulsory twenty-minute break. Amongst these songs to be orchestrated in this first session for Jamie Shaw was the planned debut single, 'When You Believe'. Ian and I had not produced any orchestra sessions since the 1980s with the Blue Mercedes album for MCA records, which was at least ten years before this session. We took the advice of the executive producer, Rick Blaskey, to use the musical arranger Richard Niles as he had been creating pop orchestral arrangements for Westlife and Boyzone. Ian and I had not met Richard Niles before but he was amiable on our approach to him and happy to record at Whitfield Street Studios with our chosen engineer Mike Ross. Richard Niles came into the P&E studio at The Strongroom for a meeting to hear the three tracks and work on the arrangements. The instructions to Richard Niles were, give us the pop Westlife-style arrangements but go towards a Disney-style orchestral arrangement on all three songs with a license to be a little more schmaltzy and bright compared with a standard BoyBand sound. Ian and I wanted harp crescendos with the strings and percussive bells from the orchestra and to make sure that the violins were high and soaring. Ian and I did not want total schmaltzy Disney, just standard pop that would suit BBC Radio 2. On the day of the session, from my perspective, two disastrous things happened that I was not prepared for:

1. The studio layout for the orchestra was in, as I soon discovered, the standard industry semi-circle (see Figure A.1). This had first and second violins to the left of the conductor and then, as the semi-circle goes around the conductor, the violas followed by the cello followed by the double basses to the right. Woodwind and harpsichord were behind them and the brass players were in separation booths (which the brass players did not really like), still within sight of the MD.
2. The arrangements that Richard Niles had written were too schmaltzy, which was potentially our fault for mentioning the word Disney to Richard, but it was as though he had ignored that Ian and I wanted to make a contemporary pop record, not a Disney show tune.

Whilst I discussed with our engineer, Mike, what could be done with the studio set-up, Ian began editing Richard's arrangement with him for the first song in an attempt to improve it. The musicians took an early break after just a couple of rehearsals for the technical levels and for Ian and I to hear the arrangement. What I wanted, technically, was to be able to pan the first violins left and the second violins right to get a rich stereo perspective of the top end of the orchestral arrangement. As Mike Ross correctly advised, this could only be done using the close microphones (which were set up) and eliminating the stereo room microphones above the conductor. To reset the studio the way I wanted it, with the first violins and second violins each side of the conductor would take Mike and his assistants over an hour. That was too much time to lose on a three-hour session where Ian and I were hoping to record orchestral arrangements on three songs. I conceded to leave the technical orchestra set up as it was and concentrated on helping Ian to edit the orchestral arrangements closer to what we considered 'contemporary pop'. Ian and I probably spent the first half hour of the session just on 'When You Believe', the track that we thought likely to be the first single from the album. Once Ian and I had achieved a good orchestra recording for 'When You Believe' we gave the orchestra another break (musician union rules). Richard Niles had a better idea of what Ian and I wanted to hear from his arrangements for Jamie's project and the editing and recording of the next two songs proceeded much quicker. Somehow we managed to finish on time and Ian and I slumped into our chairs in the control room. Once Richard and the musicians left the studio, Ian and I breathed a huge sigh of relief. Richard Niles was very patient and professional about all the changes to his arrangements and the chaos that it had caused to him and orchestral musicians, all of whom had to change their own score sheets by hand to Richard's hastily made changes.

Ian and I had already committed to commission Richard Niles to write two more orchestral arrangements for Jamie Shaw, Richard had a better idea of what was needed this time. I decided to book this second orchestral session at Angel Recording Studios in London, it was smaller than the Whitfield Street Studios and I hoped that my technical requests (this time delivered in advance) for the studio set-up and orchestra positioning would be met. Unfortunately, the Angel Studios engineer and assistant had not understood what I was attempting to achieve. They had set up the separation screens between the violins, viola and celli but they were still in the traditional classic arc or semi-circle (see Figure A.1). It took the first thirty minutes to reset the studio layout so that the first and second violins were at the front with a space in between them, then, with separating half glass acoustic screens behind them, the violas. Behind the violas, another set of screens, then the celli and double basses, more screens, then brass and woodwind with the harp in a separate booth. The results that we had achieved at Angel Studios were much better for me but I was still left the studio a little frustrated with the misunderstanding at the beginning of the session and the extra stress that it caused.

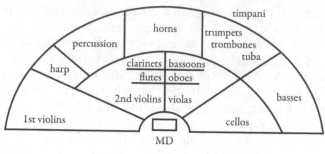

Classic Studio Set Up

Figure A.1 The 'classic' orchestra semi-circle.

Those Angel Studio sessions completed the P&E orchestral over-dubs for the Jamie Shaw album but six months later Ian and I were asked by Decca Records to produce two more tracks for Jamie in the spring of 2000, these were for another single and a repackaged version of the album. These sessions allowed Ian and I the opportunity to work with a different orchestral arranger. This time I approached Pip Williams, for whom I had assisted many years before as a junior engineer. I knew that Pip would understand my desire to re-create the sound and the studio set-up of the Marquee Studios as he had produced and arranged many orchestral overdub sessions there in the 1970s and 1980s. Pip Williams recommended that P&E book Lansdowne Studios for the orchestral session and he was also happy to come to the P&E studio for pre-production session to audition the arrangement ideas. It was during these sessions that Pip told me I was searching for 'The Beethoven Studio System' for orchestral recordings. Beethoven, the classical maestro, used this same orchestra positioning for his concerts. Figure A.2 shows an approximate copy of the Beethoven studio set-up that I have described.

My ideal set-up would have the violas and celli further back than Figure 9.2 depicts. It is interesting to note that a studio assimilation of the Beethoven Orchestra formation is my reflective perception of how an orchestra should be heard by the contemporary pop music enthusiast of the 1990s. My knowledge of this system stems from my experiences at The Marquee Studios in the 1970s and there is little commentary to be found in researching this system today[4]. I believe many creative technicians in 2019 would benefit from a basic understanding of the difference between the classic semi-circle formation compared with the little-used Beethoven Studio set-up. I do believe that the majority of modern day technologists could adapt the latter system if the opportunity arose within their recording budgets.

One of my strongest recollections of this system at The Marquee Studios was the day I assisted Marquee engineer, Geoff Calver, for the string overdubs on David Bowie's 1975 album, 'Young Americans',

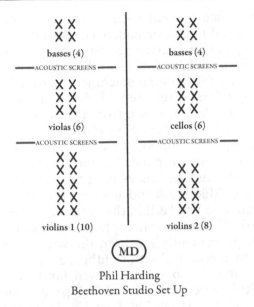

Phil Harding
Beethoven Studio Set Up

Figure A.2 The Beethoven Studio Assimilation.

with Tony Visconti producing and arranging. I remember Tony coming into the control room saying that he had arranged the strings on the assumption that we would be using our standard string set-up, which he really liked. In this particular session, Tony Visconti told Geoff Calver that there were a couple of spots in the arrangement where he had the first violins playing a part that would gradually be taken over by the second violins, so the sound would pan from left to right, across the stereo spectrum. I remember thinking that he had created this stereo soundscape out of The Marquee's studio configuration, which was generally arranged with the musicians in long, narrow banks, one behind the other.

A.4 STUDIOS: 1990S P&E MUSIC STUDIO AND STRONGROOM STUDIOS

The P&E Studio was not set up as 'just another recording studio' but more of a business enterprise, designed to be responsive to the evolving pop music of the early 1990s. It was a gamble for Ian Curnow and myself and we began with various constraints. Ian and I spent our first 6 months at The Strongroom Studio complex in a temporary programing room, situated above studio 1. Six months later P&E seized the opportunity to move into a purpose-built room called 'The Box', which was built offsite using pre-fabricated modules by specialist manufacturer, KFA, and then assembled onsite. The modules were built from 600mm-wide wood panels, pre-filled with soundproofing material.

The studio contained a small vocal booth and air-conditioning in both rooms. Ian and I also persuaded The Strongroom owner, Richard Boote, to squeeze in a machine room for our 24-track analogue tape machine, with an extra feed from the air-conditioning to keep a steady temperature for the tape machine. We were lucky enough to have natural light from a retained window that overlooked the car park and was in the line of vision from the Soundtracs mixing console. This room and two others were situated in a different building to the main Strongroom Studios. Later in the 1990s, this part of the complex would be further expanded to include The Strongroom Bar, studios 3 and 4, plus further offices and programing rooms.

An Apple Power Macintosh and an initial 4-channel Digidesign interface were purchased by P&E in the summer of 1992 and an Audio Technica AT4033 Condenser Microphone was our only microphone for eight years, permanently set up in the vocal booth. These were complimented by a powerful set of outboard equipment[5], especially the MIDI rack units armory that allowed Ian Curnow to program the typical multi-layered, orchestrated P&E productions such as 'Stay Another Day' and 'It's Alright' by East 17[6]. This level of technology gave P&E an edge over other pop and BoyBand producers of the 1990s. One of the reasons for having so many MIDI-driven sounds at Ian's fingertips was to allow us to continually monitor everything running from the computer without having to continually sync to the 2" tape machine, which would waste valuable energy and time from Ian's creative programing *flow*. Comparing the computer system to today's technology, it is surprising that it could cope with so many MIDI and audio tracks in simultaneous playback, the secret was to minimize the audio playback in the latter stages of production, sometimes I needed to sub-mix the vocal tracks (for monitoring purposes only) to allow Ian to keep working. All of this is an example of what Burgess (2015) calls 'stretching equipment and media beyond its intended use, [this] is standard practice for most producers and almost a raison d'etre for some' (Burgess, 2015, p.178). The typical final stage of a P&E production would contain just vocals as audio tracks in Cubase (plus any guitars and acoustic instruments) with all of the other instruments running live as MIDI. Once we were ready to mix, all of those parts were recorded back to 48 tracks of analogue tape in two or more passes to the Saturn analogue tape machine for my mix session in one of the main Strongroom control rooms. On the occasions when the decision was made to mix within our own studio, I would mix with the audio and MIDI tracks running live from Cubase and then record the result onto DAT[7]. In order to fulfill our commercial obligations to the record company for a typical 1990s pop project it was still necessary to deliver 2 x 24-track analogue multitrack tapes for any future external remixes that the clients may commission. In a situation that reflects the Stock and Aitken technical situation ten years before (not wishing to move away from the Linn 9000), P&E remained with the Apple/Cubase Audio/Digidesign interface system into the late 1990s even though Charlie Steinberg had demonstrated his new Cubase VST system to Ian and I during our 1998 visit to the Steinberg headquarters in Hamburg. Ian Curnow explains:

> Some would say we stuck too doggedly to that Digidesign hardware system at the P&E Studio but yes I moved straight to Cubase VST in 2000 when I set this [current at 2014] studio up.
>
> (Curnow, 2014, personal interview)

Having paid around £20,000 for the Digidesign interfaces it was difficult for Ian and I to justify moving to the new Cubase VST system, as that did not support the Digidesign hardware. Charlie Steinberg designed Cubase VST for the mass PC home-user market and we were disappointed in 1998 that he seemed to have abandoned the Cubase Audio pro-users, such as P&E, despite the extensive beta-user feedback from power users such as Ian Curnow.

A.5 P&E MUSIC STUDIO SUITE AND VOCAL BOOTH, BUILT AND EQUIPPED: 1992

Core studio essentials

1. Soundcraft Saturn 624 24-track analogue 2" tape machine + remote control.
2. Apple Quadra 700 upgraded to Power Macintosh8100 computer (updated further as the 1990s progressed).
3. Dynaudio Acoustics M2 speakers (free-standing). Powered by a stereo Pro-Studio Chord amplifier.
4. JBL control 1 nearfield monitors with the passive JBL SB1 subwoofer powered by a Quad amplifier (the speakers were placed around the dual Apple screens at the back of the room, this workstation space also contained the master Roland MIDI keyboard).
5. Soundtracs Megas MIDI 40 in-line analogue mixing console.
6. Opcode MIDI interface (Studio 4).
7. Yamaha 02R digital mixer (added in 1996). This contained an extra TC Electronic TC-Unity card AES for 02R, this added a TC m2000 FX processor to the 02R and also gave 8 x ADAT I/O's at 24-bit resolution.
8. Digidesign 442 audio interface + Digidesign PT3 card (Nubus card slot version) including expansion chassis for more Digidesign core cards.
9. Tascam DA88 + 808 remote control: An 8-track digital tape recorder (favored by the film industry as the sound and syncing was considered superior to the contemporary Alesis machine).
10. Sony PCM 2600 Professional Studio 2-channel DAT Recorder x 2.
11. Bantam rack mounted patch bay: The vital last link in the core equipment chain that was used to connect everything listed above and below.

Outboard Hardware Rack units

1. Neve (TLA conversion) rack mounting microphone pre-amp and equalizer: First in the vocal recording chain.
2. DBX 160 compressors x 2: I owned these from my days at The Marquee Studios in the 1970s. One would be last in the vocal recording chain; from here Ian and I would patch the signal directly into the Saturn analogue multitrack 2" tape machine. The other unit was used for Ian's 'Eddie' guitar sound described Appendix 3.
3. Neve 3609/C stereo compressor: This was used as a stereo master-buss compressor for rough and master mixes.
4. LA Audio Multigate: 16 x channel audio gate with MIDI control plus auto-panner, ducker, & expander.
5. Roland SDE-330 Dimensional Space Delay x 2: Classic units with delay, reverbs and RSS surround effects.
6. Roland SDE-2500 Delay x 2: Classic digital delay machines generally set to '4s' (crotchets) and '8s' (quavers).
7. Roland Dimension D SDE-320: Classic stereo chorus machine used on bass and occasionally vocals.
8. Yamaha SPX1000: Digital Reverb and effects unit.
9. Digitech TSR 24: Multi effects, reverb, delay and modulation.
10. Digitech Vocalist: Automatic vocal harmony generator, this was rarely used.
11. The Rockman: Guitar pre-amp, Ian Curnow explains; 'that was used for my mock electric guitar sounds using the Roland master MIDI keyboard and we nicknamed that 'Eddie', after Eddie Van Halen' (Curnow, 2014, personal interview).

Outboard Hardware MIDI Rack units, played from a Roland Master MIDI keyboard

1. Akai S3000 XL (x2), fully expanded with 16 Meg of RAM: These were the main samplers. They contained the majority of the P&E drum libraries and also many of our orchestral samples, including violins and 'stab hits' (orchestra samples) that were often used for dynamics into and out of choruses.
2. EMU Proteus XR Expanded: Mainly used for individual orchestral sounds such as violins, violas and celli that Ian preferred over the Akai samples, although these were often combined on the same MIDI parts.
3. EMU Vintage Keys Plus V2: Contained many piano and classic keyboard sounds such as electric piano and some synthesizer parts.
4. Kawai K5m, additive synthesis: 'Brilliant – but only for the brave!' (Curnow, 2014, personal interview). This would mainly be for unusual synthesizer sounds.

5. Korg Wavestation SR: Classically used for dance 'acoustic' piano parts plus a few other factory sounds that were useful such as organ sounds.
6. Novation Bass Station: Occasionally used for bass parts but generally combined with the SE1.
7. Novation Drum Station: Very occasionally used for drum sounds.
8. Oberhiem Matrix 1000: Regularly used for synthesizer sounds, both staccato and pads. Also some orchestral sounds were useful such as the woodwind.
9. Oberhiem OBMx 6 voice: Regularly used for synthesizer sounds, both staccato and pads. Also some orchestral sounds were useful such as the woodwind and brass.
10. Roland MKS 80 x 3: These were Ian's mainstay 'go-to' MIDI units in the early 1990s, hence there were three, used for main acoustic pianos, pads, orchestral additions such as violins.
11. Roland MPG 80: A standby back-up for the Roland MKS 80s if we wanted to have separated outputs for typical Roland pad and synthesizer sounds.
12. Roland U220: A standby back-up for the Roland MKS 80s if we wanted to have separated outputs for typical Roland pad and synthesizer sounds.
13. Roland JV880: A standby back-up for the Roland MKS 80s if we wanted to have separated outputs for typical Roland pad and synthesizer sounds.
14. Roland JD990: A standby back-up for the Roland MKS 80s if we wanted to have separated outputs for typical Roland pad and synthesizer sounds.
15. Roland JV1080 x 3 each with 4 expansion sound cards: These took over from the MKS 80s by the mid-1990s as Ian's mainstay 'go-to' MIDI units, hence there were three, used for main acoustic pianos, pads, orchestral additions such as violins.
16. Studio Electronics SE1: In Ian Curnow's words 'fantastic bass sounds, filters modeled on the Mini-Moog, with Oberhiem filters too' (Curnow, 2014, personal interview).
17. Yamaha TG77: Mainly used for typical Yamaha, clean synthesizer sounds and staccato parts.
18. Yamaha TX802: Mainly used for typical Yamaha, clean synthesizer sounds, staccato parts. This was the equivalent to 8 x Yamaha DX7's in one rack-mountable unit.

P&E Music 2nd programing room, built and equipped 1996

Core studio essentials

1. Yamaha 01 (original version) digital mixing console.
2. JBL control 1 nearfield monitors with the passive JBL SB1 subwoofer, powered with a Harman amplifier.

3. Apple Mac computer plus screen.
4. Roland Master MIDI keyboard.

Equipment rack contents

1. Akai S1000 MIDI sampler.
2. Roland JV1080 x 2 MIDI rack units.
3. Yamaha Digital reverb units x 2.

A.6 STRONGROOM STUDIOS CIRCA 1992–2000

Throughout this period. P&E Music would mainly use Strongroom Studio 2 for the final mixes described throughout this research. All of the required mix elements would be transferred back to 2 x 24 track analogue tapes as The Strongroom had not invested in the multitrack digital tape machines of this era such as the Sony PCM3324 digital machine that had existed since the late 1980s. These professional linear tape digital recorders established the 'DASH' format meaning, 'Digital Audio Stationary Head'. Sony also developed a 48-track version and the system at The Strongroom was that should clients require these formats or the 32-track Mitsubishi Digital tape machine then a hire arrangement would be required. When Ian and I arrived at The Strongroom Studio complex in early 1992, studio 2 contained an Amek Mozart mixing console that I found unsuccessful for our pop-style of mixing[8]. Owner Richard Boote purchased the Euphonix 2000 digital control surface mixing console for studio 2 in 1998. I found the sound of the Euphonix to be 'transparent', which was perfect for the pop and BoyBand multi-layered sound that Ian and I were producing. Some of our early mixes were achieved on the Neve mixing console in Studio 1 and some of our late 1990s mixes were achieved on the SSL G-Series mixing console that was housed in the newly-built Strongroom studio 3 from 1998 onwards. Here is an outline of the equipment in these studios throughout the 1990s.

Strongroom Studio 1: 1992–2000

Core studio essentials

1. Otari MTR90 24-track analogue 2" tape machines + remote control x 2 with a Lynx synchronizer system.
2. Otari ½" 2-track analogue tape machine.
3. Sony PCM 2600 Professional Studio DAT Recorder x 2.
4. Various analogue cassette tape machines.
5. Neil Grant Boxer Studio Speaker System.
6. Yamaha NS10 nearfield speakers powered by Quad amplifiers.
7. Neve VR60 Legend in-line analogue mixing console with built-in bantam patch bay.

Outboard Hardware Rack units

1. Lexicon 240: Stereo digital reverb.
2. Lexicon 480L: Dual stereo digital reverb.
3. Eventide Harmonizer.
4. Urei 1176 LN: Classic limiters.
5. Drawmer: Noise gates and compressors.
6. Roland SDE-330 Dimensional Space Delay x 2: Classic units with delay, reverbs and RSS surround effects.
7. Roland SDE-2500 Delay x 2: Classic digital delay machines generally set to 4's and 8's on my mix sessions.
8. Roland Dimension D SDE-320: Classic stereo chorus machine used on bass and occasionally vocals, this was actually our unit, permanently kept in the studio rack.
9. Yamaha SPX1000: Digital Reverb and effects unit.

Strongroom Studio 2: Late 1992–2000

Core studio essentials

Same as Studio 1 except in early 1992 the main mixing console was an Amek Mozart, replaced in late 1992 by a SSL G-Series, which was replaced around 1996 by the Euphonix 3000.

Outboard Hardware Rack units

Same as Studio 1.

Strongroom Studio 3: 1998–2000

Core studio essentials

Same as Studio 1 except the main mixing console was an SSL G-Series, which remains in situ today (2019).

Outboard Hardware Rack units

Same as Studio 1.

Notes

1. There is a full explanation of this technical process at the end of the 12-step program in Chapter 6.
2. The technical set-up of that studio is fully described in this technology appendix.
3. The full equipment list of the 1990s P&E Music Studio based at The Strongroom from 1992–2000 can be found at the end of this appendix.

4. This system is discussed at http://www.filmmusicmag.com/?p=10929 in terms of a film music library samples (accessed 23 August 2016).
5. The P&E Music Studio equipment is listed at the end of this appendix.
6. These orchestrated productions have a full commentary in Chapter 7.
7. See the end of this appendix for DAT (Digital Audio Tape) commentary.
8. An SSL G-Series replaced the Amek console late in 1992.

Glossary of Terms

A&R Record company artist and repertoire (in the 1950s and 1960s this stood for artist and recording).

Ambience The use of reverb or early reflections to create an audible sense of an acoustic space.

DAW Digital Audio Workstation

DI Direct Injection Box

DSP Digital Sound Processing

EQ Equalization

Follow-up-itis Where everyone involved from the record company A&R, manager, artists and producers tend to go around in circles chasing the track to be a perfect follow-up single to the first hit. With the same style, beats, attitude, hooks and commerciality of the first single.

Hi-Nrg Music cultural term for high-speed dance tracks that were generally associated with gay culture and gay dance clubs, 1980s–1990s.

HPF High Pass Filter

Hz The SI unit of frequency, equal to one cycle per second.

kHz A measure of frequency equivalent to 1,000 cycles per second.

L–R Left to Right (panning across the stereo soundfield).

LPF Low Pass Filter

MIDI Musical Instrument Digital Interface

'On spec' Speculative work with no advance payment, more commonly occurring in pop songwriting but occasionally may happen in pop production as well.

P&E/P&E Music Company trading name for Phil Harding and Ian Curnow.

Plots/plotted/bullseye Musically referencing a new idea to something that already exists. The terms are also used to reference instrument sounds and grooves.

PWL Company trading name for Pete Waterman and the studio complex.

SAW Company trading name for songwriters and producers Mike Stock, Matt Aitken and Pete Waterman.

VCA Voltage Control Amplifier

Bibliography

Adorno, T. (1941) On Popular Music. In *Essays on Music*. Los Angeles: University of California Press.

Adorno, T. (1970, 1997) *Aesthetic Theory*. London: Athlone Press.

Adorno, T. (2002, 1931–1969) *Essays on Music*. Los Angeles: University of California Press.

Andersson, B. (2017) Interview at Art of Record Production Conference, 2017. KMH, Royal College of Music, Stockholm: Sweden.

Bendall, H. (2015) Guest Lecture at West Suffolk College, 2015. Suffolk: England.

Bennett, J. (2015) Available at: http://joebennett.net/2015/02/27/saw-kylie-lucky/#more-4305 (accessed: 20 March 2016).

Bourdieu, P. (1984, 2010) *Distinction*. Oxford: Routledge.

Bourdieu, P. (1996) *The Rules of Art: Genesis and Structure of the Literary Field*. Cambridge: Polity Press.

Burgess, R. (1994, 1997, 2002, 2013) *The Art of Record Production*. London: Omnibus Press.

Burgess, R. (2014) *The History of Music Production*. London: Oxford University Press.

Byrne, D. (2012) *How Music Works*. Edinburgh: Canongate Books.

Cauty, J. & Drummond, B. (1988, 1998) *The Manual: How to Have a Number One the Easy Way* 2nd edn. London: Ellipsis.

Cohen, E. (1994, 2014) *Designing Groupwork: Strategies for the Heterogeneous Classroom*. New York: Teachers College Press.

Collins, M. (2009) *Pro Tools 9*. London: Focal Press.

Collins, M. (2013) *Pro Tools 11*. London: Focal Press.

Cousin, M. & Hepworth-Sawyer, R. (2014) *Logic Pro X*. London: Focal Press.

Csikszentmihalyi, M. (1975, 2000) *Beyond Boredom and Anxiety: Experiencing Flow in Work and Play*. New York: John Wiley & Sons.

Csikszentmihalyi, M. (1988) Society, Culture and Person: A Systems View of Creativity. In R. J. Sternberg (Ed.), *The Nature of Creativity: Contemporary Psychological Perspectives*. New York: Cambridge University Press.

Csikszentmihalyi, M. (1990) *Flow: The Psychology of Optimal Experience*. New York: Harper Collins.

Csikszentmihalyi, M. (1992, 2002) *Flow: The Classic Work on How to Achieve Happiness*. New York: Harper Collins.

Csikszentmihalyi, M. (1996, 1997) *Creativity: Flow and the Psychology of Discovery and Invention*. New York: Harper Collins.

Csikszentmihalyi, M. (1999) Implications of a Systems Perspective for the Study of Creativity. In R. J. Sternberg (Ed.), *Handbook of Creativity*. Cambridge: Cambridge University Press.

Egan, S. (2004) *The Guys Who Wrote 'Em: Songwriting Geniuses of Rock and Pop*. London: Askill Publishing.

Elborough, T. (2008) *The Long-Player Goodbye*. London: Sceptre Books.

Eno, B. (1979) Available at: https://techcrunch.com/2016/03/28/revisiting-brian-enos-the-studio-as-a-compositional-tool (accessed 8 April 2017).

Fauconnier, G. & Turner, M. (2002) *The Way We Think (Conceptual Blending and The Mind's Hidden Complexities)*. New York: Basic Books.

Frith, S. (1983) *Sound Effects: Youth, Leisure and the Politics of Rock'n'Roll*. New York: Pantheon Books.

Frith, S. (1996) *Performing Rites: On the Value of Popular Music*. Cambridge, Massachusetts: Harvard University Press.

Frith, S. (2011) Available at: http://pwlfromthefactoryfloor.com (accessed 30 January 2016).

Frith, S. & Zagorski-Thomas, S. (2012) *The Art of Record Production: An Introductory Reader for a New Academic Field*. Surrey: Ashgate Publishing.

Gracyk, T. (2001) *The Internet Encyclopedia of Philosophy*, ISSN 2161–0002. Available at: http://www.iep.utm.edu (accessed 29 July 2016).

Grint, B. (2019) Barry Grint (interview) chapter. In J. Hodgson and R. Hepworth-Sawyer. (Eds.), *Audio Mastering: The Artists*. New York: Routledge.

Harding, P. (2010) *PWL From the Factory Floor*. London: Cherry Red Books.

Harding, P. (2017) Top-Down Mixing: A 12-Step Mixing Program. In J. Hodgson and R. Hepworth-Sawyer. (Eds.), *Mixing Music*. New York: Routledge.

Harding, P. & Thompson, P. (2019) Collective Creativity: A 'Service' Model of Commercial Pop Music Production at PWL in the 1980s. In J. Hodgson. & R. Hepworth-Sawyer. & J. Paterson. & R. Toulson. (Eds.), *Innovation in Music: Performance, Production, Technology, and Business*. New York: Routledge.

Harding, P. & Thompson, P. (2019) Collective Creativity: A 'Service' Model of Creativity in Commercial Pop Music at P&E Studios in the 1990s. *Journal on the Art of Record Production – Proceedings of the 12th Art of Record Production Conference, 2017. KMH, Royal College of Music, Stockholm: Sweden: KMH/JARP*.

Heidegger, M. (1962,1927) *Being and Time*. Oxford: Blackwell.

Hennion, A. (1983) *The Production of Success: An Anti-Musicology of the Pop Song. Popular Music, 3*, pp.159–193. doi:10.1017/S0261143000001616.

Hepworth-Sawyer, R. & Golding, C. (2012) *What is Music Production?* London: Focal Press.

Himes, G. (2015) Available at: http://www.smithsonianmag.com/arts-culture/there-gay-aesthetic-pop-music-180956253/#BhdupiLr8b4MEQOQ.99 (accessed 30 January 2016).

Hodgson, J. & Hepworth-Sawyer, R. (2019) *Audio Mastering: The Artists*. New York: Routledge.

Husserl, E. (1927, 1971) 'Phenomenology' Edmund Husserl's Article for the Encyclopedia Britannica (1927): New Complete Translation by Richard E. Palmer, *Journal of the British Society for Phenomenology*, 2:2, pp.77–90, DOI: 10.1080/00071773.1971.11006182.

Jelbert, S. (1999) *Television Review Article*. London: The Independent Newspaper.

Jenkins, R. (1992) *Pierre Bourdieu: Key Sociologists*. London: Routledge.

Kerrigan, S. (2013) Accommodating Creative Documentary Practice Within a Revised System Model of Creativity. Newcastle, NSW: *Journal of Media Practice*. Doi: 10.1386/jmpr.14.2.111_1.

Krims, A. (2003) *What Does it Mean to Analyse Popular Music?* Music Analysis, 22: 181–209. doi:10.1111/j.0262–5245.2003.00179.x.

Lefcowitz, E. (2013) *Monkee Business: The Revolutionary Made-For-TV Band*. New York: Retrofuture Publishing.

Lefford, N. (2018) *Structuring the Creative Process, The Sonic Picture and Production Decisions – Presentation. Art of Record Production Conference. Huddersfield: UK.*

Massey, H. (1994, 2000, 2009) *Behind the Glass.* San Francisco: Backbeat Books.

Massey, H. (2015) *The Great British Recording Studios.* Milwaukee: Hal Leonard Books.

McCarthy, J. (2012) *Take That (Uncensored) on the Record.* Warwickshire: Coda Books.

McIntyre, P. (2007) *Rethinking Creativity: Record Production and the Systems Model.* Newcastle, NSW: ASARP.

McIntyre, P. (2012) *Creativity and Cultural Production: Issues for Media Practice.* Basingstoke: Palgrave MacMillan.

McIntyre, P. (2018) *Educating for Creativity within Higher Education: Integration of Research into Media Practice.* Basingstoke: Palgrave MacMillan.

McIntyre, P., Fulton, J. & Paton, E. (2016) *The Creative System in Action: Understanding Cultural Production and Practice.* Basingstoke: Palgrave MacMillan.

Moore, A. (2003) *Analyzing Popular Music.* Cambridge: Cambridge University Press.

Moore, A. (2011) Available at: http://pwlfromthefactoryfloor.com (accessed 30 January 2016).

Morey, J. (2016) The Creative Development of Sampling Composers. In P. McIntyre., J. Fulton. & E. Paton. (Eds.), *The Creative System in Action: Understanding Cultural Production and Practice.* Basingstoke: Palgrave MacMillan.

Moylan, W. (2015) *Understanding and Crafting The Mix: The Art of Recording.* Oxford: Focal Press.

Napier-Bell, S. (2001) *Black Vinyl, White Powder.* London: Ebury Press.

Negus, K. (1992) *Producing Pop: Culture and Conflict in the Popular Music Industry.* London: Oxford University Press.

Oboh, K. (2012) Available at: http://goodmenproject.com/arts/the-future-of-music-the-rise-of-technology-and-global-music *The Future of Music: The Rise of Technology and Global Music* (accessed 02 August 2016).

Parnell, M (2019) Mandy Parnell (interview) chapter. In J. Hodgson. & R. Hepworth-Sawyer. (Eds.), *Audio Mastering: The Artists.* New York: Routledge.

Perry, M. (2008) *How to be a Record Producer in the Digital Era.* New York: Billboard Books.

Sawyer, K. (1997, 2003) *Group Creativity: Music, Theatre, Collaboration.* New York: Taylor & Francis.

Sawyer, K. (2007) *Group Genius: The Creative Power of Collaboration.* New York: Basic Books.

Sawyer, K. (2013) *Zig Zag: The Surprising Path to Greater Creativity.* San Francisco: Jossey-Bass.

Scheps, A. (2018) *Keynote Speech.* Art of Record Production Conference. Huddersfield: UK.

Seabrook, J. (2015) *The Song Machine: Inside the Hit Factory.* London: Random House Publishing.

Smith, A. (1999) *Television Review Article.* London: The Guardian Newspaper.

Smith, J., Flowers, P. & Larkin, M. (2009) *Interpretative Phenomenological Analysis: Theory, Method and Research.* New York: Sage Publishing.

Stock, M. (2004) *The Hit Factory: The Stock, Aitken & Waterman Story.* London: New Holland Publishers.

Tagg, P. (2012) *Music Meanings: A Modern Musicology for Non-Musos.* New York: Mass Media Music Scholars Press.

Thompson, P. (2016) Scalability of the Creative System inside the Recording Studio. In P. McIntyre., J. Fulton. & E. Paton. (Eds.), *The Creative System in Action: Understanding Cultural Production and Practice*. Basingstoke: Palgrave MacMillan

Visconti, T. (2007) *Bowie, Bolan and the Brooklyn Boy*. London: Harper Collins.

Walsh, L. (2007) *Fast Track to Fame*. London: Bantam Press.

Waterman, P. (2000) *I Wish I Was Me: The Autobiography*. London: Virgin Books.

Watkins, T. (2016) *Let's Make Lots of Money: Secrets of a Rich, Fat, Gay, Lucky Bastard*. London: Virgin Books.

Webster, J. (1995) *Music Week Column*. London: The Music Week Magazine.

Weisberg, (1986) *Creativity: Genius and other Myths*. New York: W.H. Freeman and Company

Zagorski-Thomas, S. (2014) *The Musicology of Record Production*. Cambridge: Cambridge University Press.

Television and Radio Quotes

Arden, D. (2016) *The Music Moguls: Managers*. London: BBC4 (accessed 15 January 2016).

Brennan, L. (2016) *Who's Doing the Dishes?* London: ITV (accessed 29 January 2016).

Burns, P. (2015) *#1s of the 1980s*. London: ITV (accessed 10 November 2015).

James, A. (2018) Hits, Hype and Hustle. London: BBC4 (accessed 15 June 2018).

Levine, S. (2016) *The Music Moguls: Producers*. London: BBC4 (accessed 15 January 2016).

Napier-Bell, S. (2016) *The Music Moguls: Managers*. London: BBC4 (accessed 15 January 2016).

Discography

East 17, 'Stay Another Day' single. [Vinyl & CD]. UK: London Records, 1994.

East 17, 'Up All Night' album. [Vinyl & CD]. UK: London Records, 1996.

East 17, 'House of Love' single. [Vinyl & CD]. UK: London Records, 1992.

East 17, 'It's Alright' single. [Vinyl & CD]. UK: London Records, 1993.

Boyzone, 'Words' single. [Vinyl & CD]. UK: Polydor Records, 1996.

Bee Gees, 'Words' single. [Vinyl]. UK: Polydor Records, 1968.

Jamie Shaw, 'When You Believe' single. [CD]. UK: Decca Records, 1999.

Case Study #1 – Oslo students 2015

1. I Can't Wait.
2. Free Your Mind.
3. Stand Up.
4. You Make Me Whole Again.

Index

Note: *Italic* page numbers refer to illustrations and page numbers followed by "n" denote endnotes.